JOHN DEHART
CS 517
WASHINGTON UNIV.

Elementary combinatorial analysis

Elementary combinatorial analysis

Martin EISEN
Temple University
Philadelphia, Pennsylvania

GORDON AND BREACH
Science Publishers

NEW YORK LONDON PARIS

Gordon and Breach, Science Publishers, Inc.
One Park Avenue
New York, N.Y. 10016

Gordon and Breach Science Publishers Ltd.
42 William IV Street
London WC2N 4DF, England

Gordon & Breach
7–9 rue Emile Dubois
75014 Paris, France

First Printing October 1969
Second Printing July 1976
Third Printing September 1977

Library of Congress catalog card number 73-86631. ISBN 0 677 02260 3. Printed in the United States of America.

To
Carole, Ruth, Deborah, and Naomi

Preface

MANY STUDENTS in elementary statistics and probability courses have a great deal of difficulty solving probability problems. This is because such courses fail to provide enough illustrative examples and enough enumerative methods. The purpose of the first five chapters of this book is to remedy this deficiency.

The topics covered in the first four chapters include combinations, permutations, recursion, generating functions, principle of inclusion and exclusion, hit and rook polynomials. The fifth chapter consists of a large number of probability problems which use the methods of the previous chapter.

The book was written so that it could be understood by high school honors students or undergraduates. No mathematical background beyond high school algebra is required. The wide applicability of combinatorial methods makes the book useful for engineers, physical and social scientists, operations researchers and actuarial scientists.

Readers interested in applications can apply the results without bothering with the proofs. Although some of the proofs are not rigorous, the main ideas behind each theorem are outlined so that the interested student can learn to think for himself.

The last chapter is aimed at mathematicians and describes some current research areas. It is self contained; modern algebra is not required since all the necessary theory is presented. The methods presented in the first five chapters seem unrelated to each other. Each new theorem requires a new method of proof—there does not seem to be any underlying theory. This state of affairs is unsatisfying to a mathematician. The last chapter shows how Möbius inversion leads to the principle of inclusion and exclusion, Polya's theorem and its generalizations. This ties together most of the earlier results.

It is a pleasure to thank M. Tainiter for several suggestions.

The author is indebted to his wife, Carole, not only for her encouragement and understanding but for frequently rewriting whole sections, working out problems and editorial assistance.

July 1969 M. EISEN

Contents

CHAPTER I

Permutations and combinations

1 Introduction

WHAT IS probability theory? Intuitively, we use it every day. "Probably it will rain today—I had better wear a raincoat." "If I get 85 on this test mayby I will get an A in the course." Imagine a world with no way of knowing what to expect. Suppose it was just as likely that all 8 million inhabitants of New York would suddenly decide to cross the George Washington Bridge as that all but a reasonable number would stay home. Suppose there were no way of estimating how many babies would be born next year—how many people die—how many couples would get married and need new homes, furniture etc. There would be no way of ordering our lives—of planning ahead. We would be afraid to go out for fear of being stuck forever in a 2 million car traffic jam. The world is full of surprises but we must know what to reasonably expect if we are to plan ahead. Probability theory brings order into our lives by doing just that.

Way back in antiquity, one of the earliest prehistoric men said to himself, "Here are some fresh animal tracks, probably if I follow them I will find the animal who made them."—and so probability theory began. It was not until the 16th century, however, that anyone tried to analyze chance phenomena mathematically. Man has always been (and still is) interested in betting on dice and other games of chance. There is a great and immediate profit in being able to tell when the odds are in your favor. So it is not too surprising that the first application of mathematics to probability theory was in the field of gambling. The earliest indication that there are different probabilities for the various throws which can be made with three dice appears in a commentary (1477) on Dantes' Divine Comedy. Around 1520, Gerolamo Cardano (1501—1576) calculated the odds attached to rolling various numbers and kinds of dice. Little interest was shown in these results and it was not until 1663 that his book* *Liber de Ludo Aleae* was published.

The next investigation appears to be that of Galileo (1564—1642) who

* *See* Cardano, *The Gambling Scholar*, by Oystein Ore, Princeton University Press, Princeton (1953).

calculated the number of outcomes which occur when 3 dice are thrown. However, the founders of probability theory are generally acknowledged to be Blaise Pascal (1623—1662) and Pierre Fermat (1601—1665). In a series of letters, written around 1654, they discussed some problems raised by a gambler, the Chevalier de Mérè. The techniques and methods developed by them initiated the theory.

Most people, without any study of probability theory, would have no trouble answering the following questions: "What are the odds of getting a head when tossing a coin?" "What are the odds of getting a 6 when tossing a die?" To solve the first problem, we might say to ourselves, "I can get a head or a tail, that is, there are two possible outcomes but only one of the outcomes is the one I want; therefore the odds are one to one." Similarly, we would decide that the second answer must be 1 to 5 because out of 6 possible outcomes of tossing a die (i.e. a 1 or 2 or 3 or 4 or 5 or 6 may appear) one of them (the 6) is the one we want. However it was not until the mid eighteenth century that the following definition of probability was first clearly stated by Pierre Simon de Laplace (1749—1827):

$$\text{probability} = \frac{\text{number of favorable cases}}{\text{total number of cases}}.$$

Let us consider our two examples more closely. The rolling of a die or the tossing of a coin is called an *experiment*. Any of the outcomes is called an *event*.* Examples of events upon rolling a die are:

a) Obtaining a 6.
b) Obtaining a 3.
c) Obtaining a 1 or 3.
d) Obtaining an even number.
e) Obtaining a 7.
f) Obtaining a 1 or 2 or 3 or 4 or 5 or 6.

Examples (a) and (b) are called *simple* or *elementary* events while (c), (d), and (f) are called *compound* events. (e), which cannot occur, is called the *impossible event*. The compound event (f) which must occur is called the *sure event*. A compound event may be decomposed into two or more different elementary events (i.e. written in the form "A_1 or A_2 or $\cdots$ or A_n"). In our dice rolling experiment there are 6 simple events. ((a) and (b) are two of them.) (d) is a compound event as it may be written as obtaining a 2 or a 4 or a 6. The same

* We do not attempt to define event, just as a point in geometry is not defined. We assume the reader can recognize events. An event is just some statement about the outcomes.

event can often be described in many different ways. For example, (e) can also be written as "obtaining a -6" while (d) can be written as "obtaining a 2 or a 4 or a 6 or an 8."

Note that a compound event occurs if and only if any of the elementary events composing it occur. For example, the event "an odd number comes up" occurs if a 1 is rolled. Furthermore, if A is an event and B is an event then "A or B" is an event. For instance, if A is the event in (a) while B is the event in (b), then "A or B" is the event "obtaining a 6 or 3". More generally, if $A_1, A_2, \ldots, A_n$ are n events then "A_1 or A_2 or $\cdots$ or A_n" is an event. Analogously, if A is an event and B is an event then "A and B" is an event. For example, let A be the event "getting a number greater than 3" and B be the event described in (d). Then "A and B" is the event "obtaining a 4 or a 6". In general, if we have n events A_i $(i = 1, \ldots, n)$ then "A_1 and A_2 $\cdots$ and A_n" is an event. Finally if A is an event then so is "not A". If A is the event described in (a) then "not A" is the event "obtaining 1 or 2 or 3 or 4 or 5" while if B is the event described in (d) then "not B" is the event "obtaining an odd number".

Laplace's definition of probability can now be stated more precisely. We assume the events and experiments have the following characteristics:

a) There are only a finite number, say n (n is a positive integer), of elementary events (total number of cases).

b) The elementary events are equally likely.

c) The event A of interest can be uniquely decomposed into m elementary events (favorable cases) where $m \leqslant n$ is a nonnegative integer.

Under the above assumptions the probability of the event A, denoted by PA, is defined as

$$PA = \frac{\text{number of favorable cases}}{\text{total number of cases}} = \frac{m}{n}. \tag{1}$$

Note that according to the above definition $0 \leqslant PA \leqslant 1$ and $PA = 1$ if and only if $m = n$ i.e. if and only if A is the sure event. The impossible event is assigned a probability of zero. This is consistent with the above definition since the impossible event can be described as "none of the elementary events occur" (i.e. $m = 0$).

Example 1 If a coin is tossed and allowed to come to rest on a table, there are 2 elementary outcomes* "heads" and "tails". We assume heads

* We neglect the possibility that the coin rolls away or stands on its edge. We toss the coin over again if this occurs.

and tails are equally likely*. The probability of getting heads is $\frac{1}{2}$ according to (1) since there is only one favorable case.

Example 2 A die is rolled and allowed to come to rest on a horizontal surface. There are six elementary outcomes. The die is called fair if the six outcomes are assumed to be equally likely; otherwise it is said to be biased. Let A be the event we get a one and B be the event an odd face turns up when the die is rolled. Then $PA = 1/6$ while $PB = 3/6$ for a fair die, since the event B can be decomposed into 3 simple events.

Example 3 A pack of 52 cards, labeled from 1 to 52 is shuffled and a card is then drawn. There are 52 elementary outcomes. The phrases, the shuffling is "random" or "thorough" or the pack is "well-shuffled", will mean that we are assuming that these outcomes are equally likely. For a well-shuffled deck the probability of drawing a six or a ten or an eighteen is 3/52. The probability of choosing an even number is 26/52.

In each of the above examples we had to calculate the total number of outcomes and the favorable number of outcomes. The calculations involved were trivial. Now suppose we pose the problem of finding the probability that the uppermost faces total three when two dice are tossed. By listing the various possibilities it is not difficult to see that there are 36 possible outcomes of which 2 are favorable.

Number on green dies \ Number on red die	1	2	3	4	5	6
1	(1,1)	(1,2)	(1,3)	(1,4)	(1,5)	(1,6)
2	(2,1)	(2,2)	(2,3)	(2,4)	(2,5)	(2,6)
3	(3,1)	(3,2)	(3,3)	(3,4)	(3,5)	(3,6)
4	(4,1)	(4,2)	(4,3)	(4,4)	(4,5)	(4,6)
5	(5,1)	(5,2)	(5,3)	(5,4)	(5,5)	(5,6)
6	(6,1)	(6,2)	(6,3)	(6,4)	(6,5)	(6,6)

* Such a coin (die) is called "true", or "well-balanced". Otherwise it is called "biased". The study of the behavior of biased coins, dice etc. is beyond the scope of this book.

However, if 100 dice are tossed and we desire the probability of the upturned faces totalling 150, it is no longer practical to count the favorable cases or the total number of cases. Hence we leave the subject of probability for awhile* and concentrate on finding economical methods for enumerating the number of ways of performing some well-defined operation. This branch of mathematics, of interest in its own right, is called combinatorial analysis.

Before applying Laplace's definition, the reader must ascertain whether or not the hypotheses are satisfied. In some of the exercises at the end of this section Laplace's definition cannot be used. One might intuitively feel, in certain situations that Laplace's definition is applicable†. For example, a coin might appear to be well-balanced; however, this can only be determined by actually tossing the coin a large number of times and recording the number of heads which appear. The reader interested in biased coins and similar phenomena in which Laplace's definition does not apply is referred to Eisen, Introduction to Mathematical Probability Theory, Prentice-Hall, 1969.

Since we will only consider Laplace's definition of probability in this book we will assume from now on that each experiment has only a finite number of elementary events. In this case, each compound event A may be written as "A_1 or A_2 or $\cdots$ A_n" where n is finite and each A_i is elementary. Each A_i is called an *elementary event associated with the compound event A*. If A is a simple event, A is said to be associated with itself. The impossible event has no elementary event associated with it.

Exercises

In each of the following experiments describe the elementary outcomes, discuss whether or not they are equally likely, and state whether Laplace's definition is applicable.

a) An urn is filled with 10 identical balls numbered from 1 to 10. The balls are thoroughly mixed. A ball is selected "randomly" and its number recorded.

b) Five cards are selected from an ordinary well-shuffled deck and the number of kings obtained is recorded.

c) A university student is measured and his height is recorded to the nearest inch from 3 ft. to 7 ft.

d) A fair coin is tossed until the same result occurs twice in succession.

e) 10 well-balanced dice are rolled simultaneously and their total recorded.

---

* The reader interested in the history of probability can consult Todhunter, *A History of the Mathematical Theory of Probability from the time* of *Pascal to Laplace*, Chelsea, New York (1949).

† Laplace's definition of probability has been criticized in that it leaves "equally likely" undefined. However (b) can be omitted and (1) can be used.

f) A deck of cards numbered from 1 to 52 is thoroughly shuffled. Two cards are selected and their numbers are recorded.

g) A loaf of raisin bread is selected at random from a large number of loaves and the number of raisins in the loaf are recorded.

h) A positive integer is chosen at random.

2 Two counting principles

The first counting principle is suggested by the following problem. Suppose there are 5 roads leading from town A to town B and 3 different roads leading from town B to town C. How many different ways are there of getting from A to C? The following tree diagram suggests a method of solution.

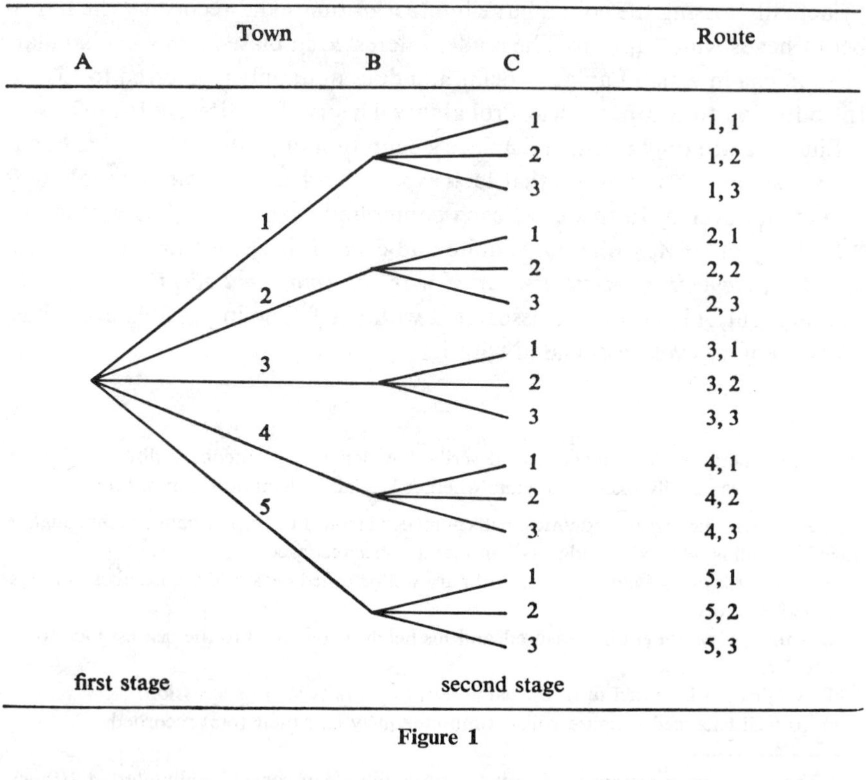

Figure 1

To get from A to B, we must start on one of the 5 roads (the branches of the tree joining A and B labelled 1 through 5) leading from A to B. Once we have chosen a road, say 1, we eventually arrive at B. Then there are 3 possible routes to C. Hence there are 3 possible routes from A to C, using road

1—namely, 1,1; 1,2; 1,3 as indicated in Fig. 1. Since each of the other branches joining A to B also has 3 branches leading from B to C there are $5 \cdot 3 = 15$ branches altogether. Hence there are 15 different ways of getting from A to C. This principle can be generalized.

THEOREM 1 *If an event A can occur in r different ways*, and for each of these ways a subsequent event B can occur in s different ways, then the event A and B can occur in rs different ways.*

Proof Drawing a tree diagram similar to figure 1, there are r branches in the first stage and s branches in the second stage and so there are rs branches altogether.

It is instructive to prove theorem 1 by means of the following alternative proof which emphasizes the role played by elementary events. Let a_1, a_2, a_3, ..., a_r represent the different ways that the first compound event A can occur and b_1, b_2, b_3, ..., b_s represent the different ways that B can occur. Consider the table below.

		Ways for B						
		b_1	b_2	b_3		b_j		b_s
Ways for A	a_1	a_1b_1	a_1b_2	a_1b_3		a_1b_j		a_1b_s
	a_2	a_2b_1	a_2b_2	a_2b_3		a_2b_j		a_2b_s
	a_i	a_ib_1	a_ib_2	a_ib_3		a_ib_j		a_ib_s
	a_r	a_rb_1	a_rb_2	a_rb_3		a_rb_j		a_rb_s

The square labelled a_ib_j is the intersection of the row labelled a_i and the column labelled b_j and represents the fact that the event A and B occurred in the following manners: first a_i happened and then b_j. For example, a_1b_3 means that the elementary event b_3 followed the elementary event a_1. Hence the number of ways that the event A and B can occur is just the same as the total number of squares—namely, rs.

* The compound event A is composed of r elementary events and when one of these occurs the event A occurs.

In some problems it is convenient to formulate the above theorem in the following manner. If one operation can be performed in r different ways and for each of these ways a second operation can be performed in s different ways then the two operations can be performed successively in rs different ways.

Example 1 If two different dice are rolled, in how many ways can they fall?

The first die can fall in 6 ways (event A) and the second in 6 ways (event B). Therefore there are $6 \times 6 = 36$ ways in which two dice can fall (event A and B). The reader having difficulty visualizing a first and a second die should think of a red and green die (see p. 4).

Example 2 How many two-lettered words, which can be meaningless, can be formed from an alphabet of 26 letters, the two letters in the word being different?

The first letter can be chosen in 26 ways (event A) and when it is chosen we can choose the second letter in 25 ways (event B). Therefore there are 26×25 or 650 ways.

COROLLARY *If the first event A_1 can occur in r_1 different ways, for each of these ways a subsequent event A_2 can occur in r_2 different ways, for each of these ways a following event A_3 can occur in r_3 different ways, ..., and for each of these ways the last event A_n can occur in r_n different ways, then the event "A_1 and A_2 and $\cdots$ and A_n" can occur in $r_1 r_2 \cdots r_n$ ways.*

Proof By theorem 1 the event A_1 and A_2 can occur in $r_1 r_2$ ways. Hence again by theorem 1 the event (A_1 and A_2) and A_3 can occur in $(r_1 r_2)\, r_3$ ways. Repeating this procedure* we see that the event "A_1 and A_2 and $\cdots$ and A_n" can occur in $r_1 r_2 \cdots r_n$ ways.

Example 3 In how many ways can 4 letters be put in 4 envelopes, one in each?

Let A_1 be the event that the first envelope is filled; the event A_1 can occur in 4 ways. Let A_2 be the event that the second envelope is filled given that the first envelope has already been filled. Since the first envelope has been filled A_2 can only occur in 3 ways. If A_2 has occurred then there are only 2 letters

* The reader familiar with mathematical induction should have no difficulty in proving this corollary rigorously.

left and the third envelope can only be filled (event A_3) in 2 ways. There is only one way of filling the last envelope (event A_4). Hence "A_1 and A_2 and $\cdots$ and A_4" can occur in $4 \times 3 \times 2 \times 1$ ways.

Example 4 How many different outcomes are possible when 100 different dice are rolled?

Let $A_1, A_2, \ldots, A_{100}$ be the events that the first, second, ..., 100th die have one of its faces uppermost. Then each A_i ($i = 1, \ldots, 100$) can occur in 6 ways since there are 6 different faces. Therefore by the corollary there are 6^{100} outcomes.

The same number of outcomes are obtained if we roll 100 different colored dice or throw a single die 100 times.

To illustrate the second counting principle consider the following problem. In how many ways can we draw a heart or a spade from an ordinary deck of cards*? Since there are 13 hearts and 13 spades we can draw a heart or a spade in $13 + 13 = 26$ different ways. This counting principle for cases in which there are alternative can be generalized.

DEFINITION 1 *Two events A and B occur* simultaneously *if and only if there exists an elementary event associated with both A and B.*

In the example on page 2 the events (a) and (d), (b) and (c); (a) and (f) occur simultaneously while the events (b) and (d) do not occur simultaneously.

THEOREM 2 *If A and B are 2 events which do not occur simultaneously and A occurs in m different ways while B occurs in n different ways then the event A or B can occur in $m + n$ different ways.*

Proof Let A be composed of the elementary events $a_1, a_2, \ldots, a_m$ and B of the elementary events $b_1, b_2, \ldots, b_n$ where each elementary event a_i is different from every elementary event b_j. A or B occurs if and only if one of the a_i ($i = 1, \ldots, m$) or one of the b_j ($j = 1, \ldots, n$) occurs and there are $m+n$ of these elementary events.

Example 5 In how many different ways can we get a seven or an eleven when we toss a pair of dice†?

---

* A deck of cards consists of 52 cards. Each card is distinguished by a face value and a suit. There are 13 face values, the numbers (ace) 1, 2, 3, $\cdots$, 10 and the pictures jack, queen, king. The four suits are called spades, clubs, hearts and diamonds. Black symbols are used to denote the first two suits and red symbols for the last two.

† Unless otherwise specified we will be interested in the total of the numbers on the uppermost faces.

Let the symbol (i, j) represent the fact that we get the number i on the first die and the number j on the second die. The various ways of rolling a seven or an eleven are: (1, 6), (6, 1), (2, 5), (5, 2), (3, 4), (4, 3), (6, 5), (5, 6). Hence there are 8 ways of getting a seven or an eleven.

The above result can be generalized; we leave the proof to the reader.

COROLLARY *If n different events A_i $(i = 1, 2, \ldots, n)$ can occur in m_i different ways and if no two of the events can occur simultaneously, then the event A_1 or A_2 or $\cdots$ or A_n can occur in $m_1 + m_2 + \cdots + m_n$ different ways.*

Frequently both counting principles must be used as is illustrated in the following example.

Example 6 How many different numbers can be formed using the digits 1, 2, 3, 4 if repetitions of digits are not allowed?

Any number k $(k = 1, 2, 3, 4)$ of digits may be used. If $k = 1$ there are 4 numbers; for $k = 2$ we obtain from theorem 1 a total of $4 \times 3 = 12$ numbers. Similarly for $k = 3$ there are $4 \times 3 \times 2 = 24$ and for $k = 4$, the total is $4 \times 3 \times 2 \times 1 = 24$. Therefore the total count is $4 + 12 + 24 + 24 = 64$.

Exercises

1 I a penny and a nickel are tossed, in how many ways can they fall?

2 a) In how many ways can two prizes be given to 30 boys, without giving both to the same boy?

b) Rework (a) under the assumption that one boy can receive both prizes?

3 In how many ways can a right-hand glove and a left-hand glove, which do not form a pair, be chosen from 10 different pairs of gloves?

4 In how many ways can a consonant and a vowel be chosen from the letters of the word "orange"?

5 a) There are 6 telegrams and 2 messenger boys. In how many different ways can the telegrams be delivered if the telegrams are distinguishable?

b) Rework (a) under the assumption that the telegrams are indistinguishable.

6 There are five cards numbered 1, 2, 3, 4, 5. How many 3-digit numbers can be formed by placing three cards in a row?

7 There are exactly four different roads from town A to town B and three from town B to town C. There are also two roads from A to C bypassing town B. How many different round trips are there from A to C and back again to A?

8 In how many ways can the following letters be arranged in a row: $a, a, a, a, a, a, a, b, c$?

9 A shoe store has 24 styles of shoes. If each style were available in twelve lengths, four widths, and six colors, how many different kinds of shoes must be kept in stock?

3 Permutations

The rules developed in the last section will now be applied to enumerate the different arrangements in which order counts, of a given number of objects. For example, suppose we have 5 different colored flags. How many different signals can be formed by hoisting 3 of the flags on a mast? Let us denote the flags by F_1, F_2, F_3, F_4, F_5 and use the symbol (F_1, F_2, F_3) to indicate that F_1 is at the top of the mast, F_2 is directly below F_1 and F_3 is below F_2 and so is lowest on the mast. Evidently the signal (F_1, F_2, F_3) can be distinguished from the signal (F_2, F_1, F_3). Hence the number of different signals is just the same as the number of ways that three places in a row can be filled by 5 distinct objects. There are 5 ways to fill the first position; when it is filled the second may be filled in 4 ways; then the third can be filled in 3 ways. Therefore by theorem 2.1 the number of signals is $5 \cdot 4 \cdot 3 = 60$.

In the above problem we had $n = 5$ things and we had to make an ordered selection of $r = 3$ of them. Each ordered triple is called a *3-permutation*. The following definition is a generalization of the above concept.

DEFINITION 1 *An* r-permutation of n things *is an ordered selection (arrangement) of* r *of them. An* n-permutation *will be simply called a permutation of* n *things.*

Some examples of r-permutations follow.

Example 1 The 2-permutations of the 4 letters a, b, c, d $(n = 4, r = 2)$ are: *ab, ba, ac, ca, ad, da, bc, cb, bd, db, cd, dc*. Since there are 4 choices for the first letter and then 3 choices for the second letter, there are $4 \cdot 3 = 12$ 2-permutations. This can easily be verified by counting.

Example 2 The 3-permutations of the 5 letters a, a, a, b, c $(n = 5, r = 3)$ are: *aaa, baa, aba, aab, abc, acb, bac, bca, cab, cba, aac, aca, caa*.

Note that nothing has been said about the nature of the objects in an r-permutation; they need not be distinct (see example 2). The meaning of ordered is that two selections are regarded as different if the order of the selection is different even when the same things are selected. For instance, in example 1 the 2-permutations *ab* and *ba* are regarded as different. Hence r-permutations may be formed in two steps; first form all possible sets of r of them and then order each of these in all possible ways. In example 1, disregarding order, there are only 6 distinct pairs of letters, namely, *ab, ac, ad, bc, bd, cd*; each of these can be arranged in 2 ways yielding 12 2-permutations altogether.

Although the objects in an r-permutation can be alike we begin by studying the simple situation in which we have n distinct objects.

THEOREM 1 *The number of r-permutations of n distinct objects, denoted by* ${}_nP_r$, *is given by*

$$ {}_nP_r = n(n-1)(n-2)\cdots(n-r+1). \tag{1} $$

Proof Since there are n distinct objects, the first of the members of an r-permutation may be chosen in n ways. This done, the second may be chosen in $n - 1$ ways since there are $n - 1$ objects left to choose from. The third can then be chosen in $n - 2$ ways and so on until the rth is chosen in $n - r + 1$ ways. By the corollary to theorem 2.1 the required number is given by (1).

COROLLARY *The number of permutations of n distinct objects is*

$$ {}_nP_n = n(n-1)(n-2)\cdots 3\cdot 2\cdot 1. \tag{2} $$

Example 3 In how many ways can a drill sergeant arrange 8 marines in a row?

The number of ways is just the same as the number of permutations of $n = 8$ distinct objects and so the required number is $8\cdot 7\cdot 6\cdot 5\cdot 4\cdot 3\cdot 2\cdot 1$ by (2).

To obtain facility in doing such problems the reader should avoid formulas and argue from first principles as follows. There are 8 choices for the first position; after the first person is chosen there are 7 people to choose from for the second position and so on until there is only 1 choice for the 8th position. By applying the corollary to theorem 2.1 we obtain the same result as above.

Example 4 In how many ways can 3 prizes be distributed to twenty competitors if each person can receive at most 1 prize?

The first prize can be given in 20 ways; when it is given, the second may be given in nineteen; then the third can be given in eighteen ways. Hence the number of ways of giving the 3 prizes is $20\times 19\times 18$ which is just ${}_{20}P_3$.

The answers to the above problems involve the products of consecutive integers. For example ${}_{10}P_3 = 10\cdot 9\cdot 8$; ${}_{10}P_5 = 10\cdot 9\cdot 8\cdot 7\cdot 6$; ${}_5P_5 = 5\cdot 4\cdot 3\cdot 2\cdot 1$. For convenience in writing such expressions we introduce the following function.

DEFINITION 2 *For every nonnegative integer n we define the function n! (read factorial n)* by*

$$n! = n(n-1)(n-2)\cdots 3\cdot 2\cdot 1 \qquad (n \neq 0).$$

$$0! = 1.$$

The values of the factorial function for the first 6 integers appears below.

$$1! = 1 \qquad 4! = 4\cdot 3! = 24$$

$$2! = 2\cdot 1 = 2 \qquad 5! = 5\cdot 4! = 120$$

$$3! = 3\cdot 2\cdot 1 = 6 \qquad 6! = 6\cdot 5! = 720.$$

It is apparent that these numbers soon become quite large. For hand calculation, tables of $n!$ and $\log n!$ can be found in most collections of mathematical tables. However nowadays most problems involving large values of $n!$ are done on computing machines. For extremely large values of n there are approximate formulas† for $n!$.

We have, upon multiplying and dividing (1) by $(n-r)!$, that

$$_nP_r = \frac{n(n-1)\cdots(n-r+1)(n-r)(n-r-1)\cdots 3\cdot 2\cdot 1}{(n-r)(n-r-1)\cdots 3\cdot 2\cdot 1} = \frac{n!}{(n-r)!}. \tag{3}$$

Example 5 How many different 3 letter words can be formed from the letters a, b, c, d, e, f, g, h which (i) include the letter e (ii) do not include the letter e?

e can be the first, second or third letter of the word. By the corollary of theorem 2.2 the total number of words which contain an e is the sum of the number of words in each of these three situations.

(i) Suppose e is the first letter of the word. Then there are 7 choices for the second letter; when this letter has been fixed there are 6 choices for the third letter. Hence there are $7\cdot 6$ words beginning with the letter e by theorem 2.1. It is not difficult to see that no matter in which position the e appears there

---

* The notation n is also used.

† See H. Robbins, "A remark on Stirlings formula", *American Mathematical Monthly*, **62**, 26–29 (1955).

are $7 \cdot 6$ different words. Hence the number of words containing an e is $3 \cdot 7 \cdot 6$.

(ii) If the letter e is not included, a word is just a 3-permutation formed from the 7 remaining letters and so there are ${}_7P_3$ words of this type.

Any 3-permutation of the above letters must either include e or not include e. Hence ${}_8P_3$ equals the number of 3-permutations including e plus the number of 3-permutations not including e and so by (i) and (ii) above

$$ {}_8P_3 = 3({}_7P_2) + {}_7P_3 . $$

The above result can be generalized. We have

$$ {}_nP_r = r\,({}_{n-1}P_{r-1}) + {}_{n-1}P_r . \tag{4} $$

This result follows from the fact that every r-permutation of the n distinct letters $a_1, a_2, \ldots, a_n$ either includes the letter a_1 or does not include the letter a_1.

${}_nP_r$ is equal to the number of r-permutations containing a_1 plus the number of r-permutations not containing a_1.

Let $A_1, A_2, \ldots, A_r$ be the events, a_1 is the 1st, 2nd, …, r-th letter of the r-permutation. Each A_i $(i = 1, 2, \ldots, r)$ can occur in ${}_{n-1}P_{r-1}$ ways since if a_1 is in the i-th position there are $r - 1$ positions to be filled from $n - 1$ distinct letters. By the corollary to theorem 2.2 the number of r-permutations containing a_1, being the event A_1 or A_2 or $\cdots$ or A_n, is just the number of ways that A_1 can occur plus the number of ways that A_2 can occur plus $\cdots$ plus the number of ways that A_r can occur. This is equal to $r({}_{n-1}P_{r-1})$. The number of r-permutations not containing a_1 is just the number of r-permutations which can be formed from the $n - 1$ remaining letters, that is, ${}_{n-1}P_r$. Adding these last two results we obtain (4).

Formula (4) can also be verified analytically by substituting the values for ${}_{n-1}P_{r-1}$ and ${}_{n-1}P_r$ given by (1) and showing that the right side reduces to the left side.

To illustrate the method of handling permutations with repeated elements consider the following example.

Example 6 How many different words can be formed by rearranging the letters of the word engineer?

Let x be the number of permutations. Consider a particular permutation, for example, *eeenginr*. Suppose the letters e were distinct, that is suppose they were written e_1, e_2, e_3, then this permutation would give rise to 3! different permutations—namely.

$$\begin{array}{ll} e_1e_2e_3nginr & e_2e_1e_3nginr \\ e_1e_3e_2nginr & e_3e_1e_2nginr \\ e_2e_3e_1nginr & e_3e_2e_1nginr. \end{array}$$

These correspond to the different ways of arranging 3 letters e_1, e_2, e_3. Since each permutation gives rise to 3! permutations with distinct e's and there are x permutations there are $3!\,x$ permutations of the letters $e_1e_2e_3nginr$. If the n's were written n_1, n_2, then the permutation $e_1e_2e_3nginr$ would give rise to 2! permutations—namely,

$$e_1e_2e_3n_1gin_2r \qquad e_1e_2e_3n_2gin_1r.$$

Reasoning as before, we see that there are $2!\,(3!\,x)$ different permutations of the letters $e_1e_2e_3n_1gin_2r$. Since the letters are now all distinct this number must be equal to 8!, that is, $2!\,(3!\,x) = 8!$ or $x = \dfrac{8!}{2!\,3!}$.

In the above example there are $n = 8$ different letters, $k_1 = 3$ letters of the first kind (e's), $k_2 = 2$ letters of the second kind (n's), $k_3 = 1$ letter of the third kind (g), $k_4 = 1$ letter of the fourth kind (i), and $k_5 = 1$ letter of the fifth kind. We now turn to the general situation.

THEOREM 2 *The number of permutations of n things when there are k_1 like things of one kind, k_2 like things of another kind, ..., k_r like things of the r-th kind, and $k_1 + k_2 + \cdots + k_r = n$ is*

$$\frac{n!}{k_1!k_2! \ldots k_r!}. \tag{5}$$

Proof Let x be the number of permutations. If the k_1 like things were all different there would be $(k_1!)\,x$ permutations since each old permutation would give rise to $k_1!$ different permutations corresponding to the number of ways of arranging k_1 distinct objects in a row. If the k_2 like things were now changed into distinct objects there would be $k_2!\,(k_1!\,x)$ permutations by similar reasoning. Repeating this procedure until all the objects are distinct will give $(k_r!) \cdots (k_3!)\,(k_2!)\,(k_1!)\,x$ permutations. However by the corollary to theorem 1 we know that there are $n!$ permutations of n distinct objects. Equating these two quantities and solving for x yields (5).

Example 7 In how many ways can three 6's and two 5's be obtained when five dice are cast?

This problem is equivalent to the number of ways of arranging three 6's and two 5's in a row and so the answer is $\dfrac{5!}{3!\,2!}$.

In general, a simple formula analogous to (1) for r-permutations of n objects which are not distinct does not exist. However we shall see in the following section, that such problems can be reduced to applications of theorem 2.

Some permutation problems have restrictions whereby certain conditions are imposed. The following examples illustrate how to handle such problems.

Example 8 How many distinct 3 digit numbers are there which are even and have no repeated digits?

For a number to be even it must end in 0, 2, 4, 6, or 8. There are two situations to consider. First suppose that the number ends in 0; then there are 9 choices for the first digit of the number and once this is fixed there are 8 choices for the second digit. Hence there are $9 \cdot 8$ numbers which end in a zero. Now suppose the number does not end in zero. Then there are 4 choices for the last position (2, 4, 6, 8); when this number has been chosen there are only 8 choices for the first digit, since the number cannot begin with zero. Finally there are 8 choices for the second digit and so there are $8 \cdot 8 \cdot 4$ numbers which do not end in zero. Accordingly there are $9 \cdot 8 + 8 \cdot 8 \cdot 4$ even 3-digit numbers with no repeated digits.

Example 9 In how many ways can 10 men be arranged in a row given that 3 particular men must always stand next to each other?

Since the three men always stand next to each other they can be thought of as a single object, which will be denoted by X. Let M_i ($i = 1, \ldots, 7$) denote the remaining men. We first find the number of ways of arranging the objects $X\, M_1 M_2 M_3 M_4 M_5 M_6 M_7$. By the corollary to theorem 1 this can be done in $8!$ ways. Each permutation of the above type gives rise to $3!$ permutations due to the fact that the three men represented by X can be permuted among themselves. Hence there are $(3!)\,(8!)$ permutations altogether*.

Care must be exercised in forming circular arrangements. The next three examples illustrate different ways in which such arrangements can be interpreted.

* This example might be interpreted to mean the specific three men have a particular arrangement. The answer would then be $8!$

Example 10 A circular table is laid for 7 people. In how many ways can they be seated if each chair is different?

By the corollary to theorem 1 the answer is 7!

Example 11 In how many ways can 7 children form themselves in a ring, to play ring-around-the-rosie?

Here the 7 children are not assigned to particular places but are only arranged relative to one another. That is, the arrangements

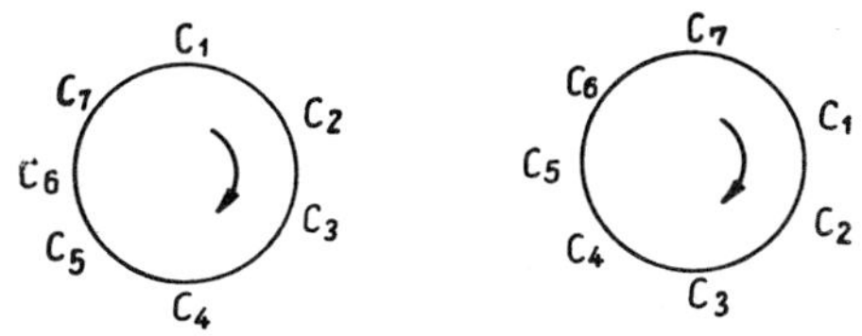

are considered the same if the children are in the same order clockwise. In other words, if we start with child C_1 and go around the circle clockwise the children are reached in the same order. Hence the position of child C_1 is immaterial and it is only the position of the six other children relative to C_1 which counts. Therefore keeping C_1 fixed in position there are 6! arrangements of the remaining children by the corollary to theorem 1. This is the required result.

Example 12 In how many ways can seven different colored spherical beads be strung on a string to form a necklace?

Here, unlike example 10, no distinction is made between clockwise and counterclockwise arrangements. If the necklace is turned over each clockwise arrangement corresponds to a unique counterclockwise arrangement which must be regarded as equivalent. For example, the two arrangement shown below are equivalent:

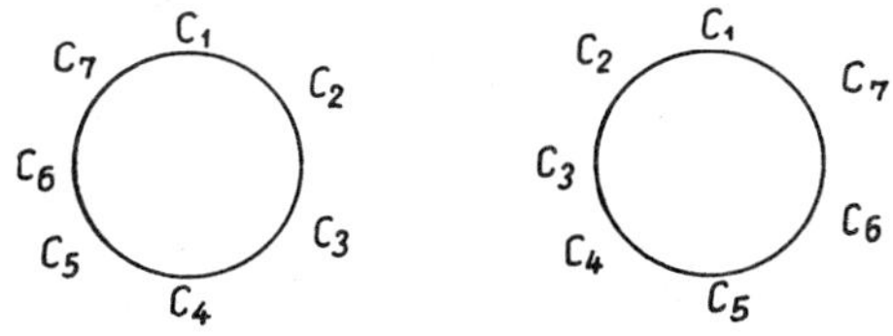

Hence there are $\dfrac{(7-1)!}{2}$ different necklaces.

Finally let us consider r-permutations when there are n kinds of things but an unlimited number of each kind. Such a situation can be created by replacing any chosen object by a similar object in stock.

Example 13 A small hotel offers fish, steak, chicken or roast beef for dinner. How many different sets of dinners may one person order in a week?

There are 4 choices on each day of the week. Therefore the number of different arrangements for a week is $4 \cdot 4 \cdot 4 \cdot 4 \cdot 4 \cdot 4 \cdot 4 = 4^7$.

This result can be generalized.

THEOREM 3 *The number of r-permutations of n unlike things, if each may be repeated any number of times, is n^r.*

Proof Each position in an r-permutation can be filled in n ways and so by the corollary to theorem 2.1 the number of r-permutations of n unlike thing with repetitions is n^r.

Exercises

1 How many marriageable couples can be formed from 6 boys and 9 girls?

2 In how many ways can 1 novel, 1 magazine and 1 newspaper be chosen from 10 novels, 8 magazines, and 5 newspapers?

3 Every day 8 people sit on a long bench to eat breakfast. They decide to sit in a different order every day. When will they have exhausted all possibilities if they begin on Jan. 1, 1954?

4 In how many ways can 5 boys and 4 girls be seated in a row so that every girl sits between 2 boys?

5 In how many ways can the letters of the word "Mississippi" be arranged? How many arrangements begin with m? How many begin with m and end in i?

6 In how many ways can 10 people be arranged at a round table so that 2 particular people sit together?

7 There are m stations on a railroad line. How many different types of tickets must be printed if it is possible to book from any one station to any other?

8 Express in factorials:

a) $12 \times 11 \times 10$
b) $7 \times 8 \times 9 \times 10$
c) $n(n^2 - 1)(n^2 - 4)(n^2 - 9)$
d) $n(n + 1)(n + 2) \cdots (n + r)$.

9 Verify the following identities:

a) $(2n)! \div n! = 2(1 \cdot 3 \cdot 5 \cdot 7 \cdots (2n - 1))$.
b) $2 \cdot 6 \cdot 10 \cdot 14 \cdots (4n - 6)(4n - 2) = (n + 1)(n + 2)(n + 3) \cdots (2n - 1) 2n$.

10 A shelf contains 5 mathematics books, 6 physics books, and 8 English books. In how many ways can the nineteen books be arranged, keeping all books of the same sort together?

11 In how many ways can a four-letter word be made from an alphabet of 26 letters if:

a) repetitions are not allowed?
b) repetitions are allowed?

12 a) How many numbers are there with 6 digits or less?
b) How many different symbols can be formed using only dots and dashes if at most a row of 4 of them are used?

13 Suppose every day is described as sunny, rain, or cloudy. Within how many years must there be a repetition of the description of a week's weather?

14 A zoologist is studying patterns of children in families. A family type is designated by a code—for example, MFF denotes a family of three children of which the oldest is a male and the other two females. Note that MFF, FMF, and FFM are different types. How many family types are there with at least two but not more than five children?

4 Combinations

In many situations we are concerned with the number of selections which can be made from a given set of distinguishable objects. The following example illustrates the fact that the order of the objects in a selected group can be immaterial.

Example 1 How many different games of ping-pong can be arranged if there are 5 players present?

Denoting the players by P_i ($i = 1, 2, \ldots, 5$) and the fact that P_1 plays P_2 by (P_1, P_2) we can list all the possible games*:

$$(P_1, P_2), (P_1, P_3), (P_1, P_4), (P_1, P_5), (P_2, P_3), (P_2, P_4), (P_2, P_5), (P_3, P_4), (P_3, P_5), (P_4, P_5).$$

Note that each combination yields exactly 2! permutations corresponding to the permutations of the 2 distinct objects in each selection. By the remarks following example 3.2 each 2-permutation can be formed by first selecting all possible pairs and then forming permutations of each pair. Since there are $5 \cdot 4$ 2-permutations there must be $\frac{5 \cdot 4}{2!}$ selections of 2 objects. This can be verified directly by counting.

* It is assumed that (P_1, P_2) is not different from (P_2, P_1). However, the first named player might serve first. In this situation (P_1, P_2) is different from (P_2, P_1) and the solution to the problem is 5·4.

A selection of the form (P_1, P_2) is called a 2-combination $(r = 2)$ from 5 $(n = 5)$ distinct objects. This definition can be generalized.

DEFINITION 1 *An* r-combination of n things *is a selection of* r *of them without regard to order.*

The n objects need not be distinct. For example, the 3-combinations of the objects a, a, a, a, b, b, c, c are:

$$(a, a, a), (a, a, b), (a, b, b). (a, a, c), (a, c, c), (a, b, c), (b, c, c), (b, b, c).$$

Following the procedure used in example 1 we can find the number of r-permutations of n distinct things.

THEOREM 1 *The number of r-combinations of n distinct things, denoted by* ${}_nC_r$ *or* $\binom{n}{r}$ *is given by*

$${}_nC_r = \frac{{}_nP_r}{r!} = \frac{n(n-1)(n-2)\cdots(n-r+1)}{r!} = \frac{n!}{(n-r)!\,r!}. \tag{1}$$

Proof Each r-combination of n distinct things can be ordered in $r!$ ways. Every r-permutation is obtained if we first form all possible r-combinations and then form all possible permutations of the objects in each r-combination Hence $(r!)\,{}_nC_r = {}_nP_r$ which when solved for ${}_nC_r$ yields (1) by equation 3.3.

Example 2 In how many ways can a hand of 4 cards be selected from 52 cards?

Since the order of cards in a hand does not count there are $\binom{52}{4}$ hands.

Example 3 Seven generals store top-secret plans in a safe. They want to be able to open the safe only when a majority of the group is present. They propose locking the safe with a number of different locks. Each person will be given keys to certain of these locks. How many locks are required and how many keys must each general have?

Given any three members of the group, there must be a lock for which none of them has a key. Suppose the same lock L is associated with the groups G_1, G_2, G_3, and G_1, G_2, G_4. Since G_1, G_2, G_3, G_4 form a majority, G_4 must have a key to L, which is a contradiction. Therefore the number of locks must be at least equal to the number of ways in which 3 people can be selected from 7, that is, $\binom{7}{3} = 35$.

Let G be one of the generals. We have just seen that given any set of 3 generals selected from the remaining 6, there is a lock which they cannot open. Since G has a key to this lock, G must have at least $\binom{6}{3} = 20$ keys.

These lower bounds actually work as the reader can verify.

The following example illustrates a method of calculating the number of r-permutations of n objects which need not be distinct.

Example 4 How many different 5 letters words can be formed from the letters *aaaaa, bbb, cc, ddd, ee, f, gggg*?

The different ways of selecting 5 letters may be classified as follows:

(1) all 5 alike,
(2) 4 alike and 1 different,
(3) 3 alike and 2 others alike,
(4) 3 alike and 2 others different,
(5) 2 alike, 2 others alike, and 1 different,
(6) 2 alike and 3 different,
(7) all five different.

Selection (1) can be made in only 1 way (namely by selecting a, a, a, a, a), and when this selection of letters is made they can be arranged in only 1 order. Therefore (1) gives rise to only 1 arrangement.

Selection (2) can be made in 12 ways; the 4 alike can be selected in 2 ways (viz. *aaaa* or *gggg*) and the 1 different in 6 ways. When a selection of letters has been made they can be arranged in $\dfrac{5!}{4!\,1!} = 5$ ways by theorem 3.2. Hence (2) gives rise to $12 \times 5 = 60$ arrangements.

Selection (3) can be made in 20 ways. The 3 alike can be chosen from the a's, b's, d's, or g's, and once one of these is chosen there are 5 choices for the 2 alike (since the c's and e's can now be chosen). When a selection of letters has been made they can be arranged in $\dfrac{5!}{3!\,2!} = 10$ ways. Therefore (3) yields 200 ways.

Selection (4) can be made in $4\binom{6}{2}$ ways since there are 4 choices for the 3 alike letters and once these have been chosen there are 6 different letters from which 2 must be chosen. The selection can be arranged in $\dfrac{5!}{3!\,1!\,1!}$ ways. Hence (4) gives rise to 1200 ways.

Selection (5) can be made in $\binom{6}{2}$ (5) ways since there are $\binom{6}{2}$ choices for the two 2 alike and 5 choices for the 1 letter since f can now be included. A word of this type can be rearranged in $5!/(2!\,2!\,1!)$ ways. Therefore (5) gives rise to:

$$[75]\left[\frac{5!}{2!\,2!\,1!}\right] = 2250 \text{ ways.}$$

Similar arguments show that (6) and (7) yield 7200 and 2520 words respectively. Hence the total number of words is:

$$1 + 60 + 200 + 1200 + 2250 + 7200 + 2520 = 13{,}431.$$

In general there is no simple expression for the number of r-combinations of n objects which need not be distinct.

Some numerical values for ${}_nC_r$ appear below. This array of numbers is called *Pascal's triangle*.

$$\begin{array}{ccccccccccc} &&&&& {}_0C_0 &&&&& \\ &&&& {}_1C_0 && {}_1C_1 &&&& \\ &&& {}_2C_0 && {}_2C_1 && {}_2C_2 &&& \\ && {}_3C_0 && {}_3C_1 && {}_3C_2 && {}_3C_3 && \\ & {}_4C_0 && {}_4C_1 && {}_4C_2 && {}_4C_3 && {}_4C_4 & \\ {}_5C_0 && {}_5C_1 && {}_5C_2 && {}_5C_3 && {}_5C_4 && {}_5C_5 \end{array} \qquad \begin{array}{ccccccccccc} &&&&& 1 &&&&& \\ &&&& 1 && 1 &&&& \\ &&& 1 && 2 && 1 &&& \\ && 1 && 3 && 3 && 1 && \\ & 1 && 4 && 6 && 4 && 1 & \\ 1 && 5 && 10 && 10 && 5 && 1 \end{array}$$

The observant reader will notice certain regularities in the above table. The numbers equidistant from each end in a given line are equal. For example, ${}_5C_1 = {}_5C_4$, ${}_5C_2 = {}_5C_3$, ${}_3C_1 = {}_3C_2$ etc. This leads to the following conjecture.

$${}_nC_r = {}_nC_{n-r}. \tag{2}$$

This result can be verified verbally. Observe that every time an r-combination is formed there are $n - r$ objects left. These objects can be considered as an $(n - r)$-combination. Hence to every r-combination we can associate an $(n - r)$-combination and conversely. In other words there are as many r-combinations as $(n - r)$-combinations which is just what (2) states.

Alternatively, ${}_nC_r = \dfrac{n!}{(n-r)!\,r!}$ while ${}_nC_{n-r} = \dfrac{n!}{(n-(n-r))!\,(n-r)!}$ $= \dfrac{n!}{(n-r)!\,r!}$ and so (2) is valid.

Another observation is that any number in the table (except the 1's) is the sum of the two numbers appearing to the right and left in the line above. For example, ${}_5C_1 = {}_4C_0 + {}_4C_1$, ${}_5C_3 = {}_4C_2 + {}_4C_3$, and ${}_4C_2 = {}_3C_1 + {}_3C_2$. We conjecture that

$$\binom{n}{r} = \binom{n-1}{r-1} + \binom{n-1}{r}. \tag{3}$$

$\binom{n}{r}$ is just the number of r-combinations formed from the n distinct letters $a_1, a_2, \ldots, a_n$. Every r-combination either includes the letter a_1 or does not include the letter a_1. The number of r-combinations including the letter a_1 is $\binom{n-1}{r-1}$ since there are only $n - 1$ letters to choose from and we must choose $r - 1$ letters in addition to a_1 to form an r-combination. The number of r-combinations which do not include a_1 is $\binom{n-1}{r}$ since we must choose r letters from the $n - 1$ letters $a_2, \ldots, a_n$.

Alternatively,

$$\begin{aligned}\binom{n-1}{r-1} + \binom{n-1}{r} &= \frac{(n-1)!}{(n-r)!\,(r-1)!} + \frac{(n-1)!}{(n-r-1)!\,r!} \\ &= \frac{(n-1)!}{(n-r-1)!\,(r-1)!}\left[\frac{1}{n-r} + \frac{1}{r}\right] \\ &= \frac{(n-1)!}{(n-r-1)!\,(r-1)!}\left(\frac{n}{r\,(n-r)}\right) \\ &= \frac{n!}{(n-r)!\,r!} = \binom{n}{r}.\end{aligned}$$

There may be certain restrictions imposed upon a selection. Such problems can be treated as in the following example.

Example 5 In how many ways can a committee of 3 teachers and 4 students be chosen from 9 teachers and 6 students? What is the number of ways if teacher A refuses to serve if student B is a member?

(a) There are ${}_9C_3$ ways of selecting 3 teachers from 9 teachers and ${}_6C_4$ ways of selecting 4 students from 6 students. Therefore by theorem 2.1 the number of ways of choosing the committee is $\binom{9}{3}\binom{6}{4}$.

(b) If A and B are members there remains to be selected 2 teachers from 8 teachers, and 3 students from 5 students. This can be done in $\binom{8}{2}\binom{5}{3}$ ways. Therefore the number of ways of making selections which do not include A and B is:

$$\binom{9}{3}\binom{6}{4} - \binom{8}{2}\binom{5}{3}.$$

In the following example we consider r-combinations in which repetitions are allowed.

Example 6 A store sells six flavors of ice cream. In how many ways can a customer choose 3 ice cream cones not necessarily of different flavors?

There are 6 ways in which all flavors are the same, $6 \cdot 5$ ways in which 2 flavors are the same and $(6 \cdot 5 \cdot 4)/3!$ ways in which all flavors differ. Hence the number of ways is $6 + 30 + 20 = 56 = {}_8C_3$.

The above approach is difficult to generalize—try it for 100 flavors and 25 cones. An alternative method follows.

Let us denote the different flavors by the integers 1, 2, 3, ..., 6. Then combinations such as (1, 2, 3), (1, 1, 2), and (3, 3, 3) can occur. In general we will have (c_1, c_2, c_3) where each c_i is one of the integers 1, ..., 6 and $c_1 \leqslant c_2 \leqslant c_3$. There is no loss in generality in this assumption, since in a selection the order of the numbers is immaterial. To each 3-combination (c_1, c_2, c_3) we associate a unique 3-combination (d_1, d_2, d_3) by setting $d_1 = c_1$, $d_2 = c_2 + 1$, and $d_3 = c_3 + 2$. For example for the above 3-combinations we obtain the following associated 3-combinations:

$$(1, 3, 5), (1, 2, 4), (3, 4, 5).$$

Conversely, given (d_1, d_2, d_3) we can find (c_1, c_2, c_3). Hence the problem is reduced to finding the number of 3-combinations of type (d_1, d_2, d_3). Observe that the numbers d_1, d_2, d_3 are always distinct and each d_i is one of the integers from 1 through 8. There are ${}_8C_3$ selections of ice cream cones.

The general problem is treated in a similar manner.

THEOREM 2 *The number of r-combinations of n distinct things in which repetitions are allowed is* $\binom{n+r-1}{r}$.

Proof We may assume, without loss of generality, that the n distinct things are the integers 1, 2, ..., n. Let $(c_1, c_2, \ldots, c_r)$ be an r-combination of the above type with the c_i arranged in nondecreasing order. With each r-combination associate another r-combination $(d_1, d_2, \ldots, d_r)$ by the rules:

$$d_1 = c_1 + 0, d_2 = c_2 + 1, d_3 = c_3 + 2, \ldots, d_r = c_r + r - 1.$$

Note that whatever the c's are, the d's are unlike. The d's uniquely determine the c's and the c's uniquely determine the d's. Hence each r-combination of the c's corresponds to a unique r-combination of the d's and vice versa. Therefore the two types of r-combinations are equinumerous. However the r-combinations of the d's are just selections of r integers from the integers 1 through $n + r - 1$. There are $\binom{n + r - 1}{r}$ such r-combinations.

Example 7 If a candy factory manufactures 10 different kinds of chocolate and puts them in boxes of 32, how many different boxes can be formed?

Since the order of the chocolates in a box is immaterial we are interested in finding the number of 32-combinations of 10 distinct things with repetitions allowed. By the theorem the answer is $\binom{41}{32}$

The various products of degree r which can be formed with any (0 up to n) of n given letters are called *homogeneous products of r dimensions* formed from the n letters (repetitions are allowed). For example, let $n = 4$ and $r = 3$; then the homogeneous products of 3 dimensions formed from the letters a, b, c, d are:

$$a^3, b^3, c^3, d^3, a^2b, a^2c, a^2d, b^2a, b^2c, b^2d, c^2d, \ldots, abc, abd, bcd.$$

Each of the above homogeneous products corresponds uniquely to a 3-combination as described in theorem 2; for example, $a^3 \leftrightarrow (a, a, a,)$ $b^2c \leftrightarrow (b, b, c)$ and $bcd \leftrightarrow (b, c, d)$. Hence the number of homogeneous products is just $\binom{6}{3}$ according to theorem 2. A similar result holds for homogeneous products of r-dimensions.

Example 8 In how many ways can 6 similar balls be distributed in 10 different boxes?

Let B_i $(i = 1, \ldots, 10)$ denote the boxes. Some distributions of balls in boxes are shown below.

B_1	B_2	B_3	B_4	B_5	$\cdots$	B_{10}	code
1	3	2	0	0		0	$b_1^1\ b_2^3\ b_3^2\ b_4^0\ b_5^0\ b_6^0\ b_7^0\ b_8^0\ b_9^0\ b_{10}^0$
1	1	1	2	1		0	$b_1^1\ b_2^1\ b_3^1\ b_4^2\ b_5^1\ b_6^0\ b_7^0\ b_8^0\ b_9^0\ b_{10}^0$
0	1	0	1	2		2	$b_1^0\ b_2^1\ b_3^0\ b_4^1\ b_5^2\ b_6^0\ b_7^0\ b_8^0\ b_9^0\ b_{10}^2$

The numbers appearing under the B's denote the number of balls in those boxes. For example, the first line denotes the fact that there is 1 ball in box B_1, there are 3 and 2 balls respectively in boxes B_2 and B_3 and the remaining boxes are empty. To specify a distribution of balls we adopt the following convention: if box B_i contains j_i balls we write $b_i^{j_i}$ $(i = 1, 2, \ldots, 10)$ then multiply all of the b's together. Some examples of the code appear above. Hence a placement of balls into boxes uniquely corresponds to a homogeneous product of degree 6 formed from the letters b_i $(i = 1, \ldots, 10)$. Conversely any such homogeneous product uniquely corresponds to a distribution of balls in boxes. Hence the number of ways of distributing 6 similar balls in 10 different boxes is $\binom{15}{6}$, which is just the number of homogeneous products.

Another method of attacking problems of this type depend upon the following artifice. Let the balls be represented by 6 dots and the boxes by using vertical strokes as follows. The spaces between any two strokes as well as the space before the first stroke and after the last stroke represent the boxes. The distribution of balls into boxes can then be represented by configurations such as

$$\underset{A}{|\ |\cdot|\cdot|\cdot|\cdot|\cdot|\cdot|\ |} \qquad\qquad \underset{B}{|\ |\ |\ |\cdots|\cdot|\cdot\cdot|\ |\ |}$$

which correspond to

Box		1	2	3	4	5	6	7	8	9	10
number of balls in the box	A	0	0	1	1	1	1	1	1	0	0
	B	0	0	0	0	3	1	2	0	0	0

It can be seen that each placement of the balls in the boxes can be uniquely represented by writing 9 strokes and 6 dots in a row in a certain order and conversely. Hence the number of ways of placing 6 similar balls in 10 different boxes is equal to the number of ways that 6 dots and 9 strokes can be arranged in a row. This can be done in $\binom{15}{6}$ ways by theorem 3.2.

Example 9 Show that n similar balls can be placed in r different boxes ($n \geqslant r$), so that no box remains empty, in $\binom{n-1}{r-1}$ ways.

As in example 8 the balls can be represented by n dots and the boxes by using $r - 1$ vertical strokes. Since no box is to be empty the vertical strokes must appear in the spaces between the dots. There are $n - 1$ such spaces and $r - 1$ of these must be chosen for the strokes. Therefore the number of allowable configurations of strokes and dots is $\binom{n-1}{r-1}$.

Further results of this type can be found in chapter III.

The following result will be useful in chapter IV.

THEOREM 3 *The number of r-combinations of the first n natural numbers on a line, with no two consecutive integers in the same combination is* $\binom{n-r+1}{r}$

Proof Arrange the first n natural numbers on a line according to increasing magnitude. Place a 1 under each integer selected and a 0 under each integer not selected. Then each arrangement of $p = r$ ones and $q = n - r$ zeroes corresponds to a unique r-combination and conversely. From this point of view, the theorem is equivalent to finding the number of ways p ones and q zeroes, $q \geqslant p - 1$, can be placed on a line with no two ones together.

To solve this problem, place the p ones and $p - 1$ zeroes on a line with a zero between each pair of ones. Distribute the remaining $q - (p - 1)$ zeroes in the $p + 1$ cells formed by the p ones (c.f. example 8). This can be done in

$$\binom{q-(p-1)+p}{p} = \binom{q+1}{p} \tag{4}$$

ways. This is the required result when $p = r$ and $q = n - r$.

COROLLARY *The number of r-combinations of the first n natural numbers if arranged on a circle and if no two consecutive integers are in the same combination and if n and* 1 *are consecutive is* $\dfrac{n}{n-r}\binom{n-r}{r}$.

Proof This number is the same as the number of arrangements of p ones and q zeroes on a line with no two consecutive ones and such that the first and last digits are not both ones. An allowable arrangement can occur in only 2 mutually exclusive ways.

a) The arrangement begins with a one.

The one must be followed by a zero and the last digit must be zero. Hence we need only find the number of ways $p - 1$ ones and $q - 2$ zeroes can be placed on a line with no two ones together. By (4), this number is

$$\binom{q-1}{p-1} = \frac{p}{q}\binom{q}{p}.$$

b) The arrangement begins with a zero.

The number of such arrangements is obtained from (4) with q replaced by $q - 1$, that is, $\binom{q}{p}$.

Thus, the total number of arrangements is

$$\frac{p+q}{q}\binom{q}{p}. \tag{5}$$

Setting $p = r$ and $q = n - r$ in (5) yields the desired result.

Exercises

1 There are 10 houses in a row. In how many ways can 4 of the front doors be chosen for painting yellow?

2 There are 15 letters but only enough stamps for 11 of them. In how many ways can the letters to be stamped be chosen?

3 What is the greatest number of points of intersection of (a) 10 straight lines? (b) 6 circles? (c) 7 straight lines and 4 circles?

4 In how many ways can a team of 5 be chosen from 10 players (a) so as to include both the strongest and weakest player, (b) so as to exclude the weakest and include the strongest?

5 (a) n points are marked on a circle. How many chords can be obtained by joining them in pairs? (b) How many diagonals does an n-sided polygon have, if a diagonal means any line joining two non-consecutive corners?

6 A menu in a Chinese restaurant offers a meal consisting of 5 choices from column A and 3 choices from column B. Suppose column A lists 15 different foods and column B

lists 10 different foods none of which appear in column A. How many different meals are possible?

7 The results of 25 football games (win, lose or draw) are to be predicted. How many different forecasts can contain exactly 18 correct results?

8 How many dominoes are there in a set numbered from double blank to double six?

9 In how many ways can three glasses be filled without mixing from five kinds of wine if (a) the glasses are similar? (b) the glasses are distinct?

10 In how many ways can three boys divide 12 different sized apples (a) if each takes four? (b) if the largest boy gets 6 and the other two boys get 3 apiece?

11 In how many different ways can r particles be distributed in n different cells under the assumptions:

a) (Boltzmann statistics)* All particles are distinguishable and any number of them can be in any cell?

b) (Bose–Einstein) The particles are indistinguishable and any nunber of them can be in any cell?

c) (Fermi–Dirac) The particles are indistinguishable and not more than 1 particle may be found in a cell?

12 Examine Pascal's triangle and try to formulate and prove a theorem concerning the largest value of ${}_nC_r$ for a fixed n and an arbitrary r.

13 Suppose an urn contains n balls numbered from 1 to n, k balls are drawn and the number on each ball is recorded.

a) (ordered sampling with replacement) If each ball is replaced how many different sequences of numbers will be recorded?

b) (ordered sampling without replacement) If each ball is not replaced how many different sequences of numbers will be recorded?

c) Answer (a) and (b) if the order of the numbers in the sequence is ignored (unordered sampling)?

14 There are 70 unescorted men and 65 unescorted women at a dance. Suppose we are concerned only with whether or not various men and women pair off during the evening. Assume that each person is involved in only one such pairing. (a) In how many ways can precisely k couples be formed, $0 \leqslant k \leqslant 65$? (b) In how many ways can the evening end?

15 a) A man walks fifteen blocks to his office which is located seven blocks west and eight blocks south of his home. The streets are laid out in a rectangular pattern and all the streets are available for walking. How many different paths are there from his home to his office provided that he walks only fifteen blocks?

b) Suppose that the north–south streets are labelled $A, B, C, \ldots, H$ and the east–west streets 1st, 2nd, …, 9th. His home is at 1st and A and his office is at 9th and H. Repeat (a) under the assumption that E street from 5th to 6th is closed for walking.

---

* The word "statistics" refers to the distribution of particles in cells as used in physics. For details, see F. W. Sears, *Thermodynamics*, Addison-Wesley, Cambridge, Mass. (1950).

16 The following numbers, called *Fibonacci numbers*, are obtained by rewriting Pascal's triangle in the form indicated below (the first 1 is omitted) and summing diagonally.

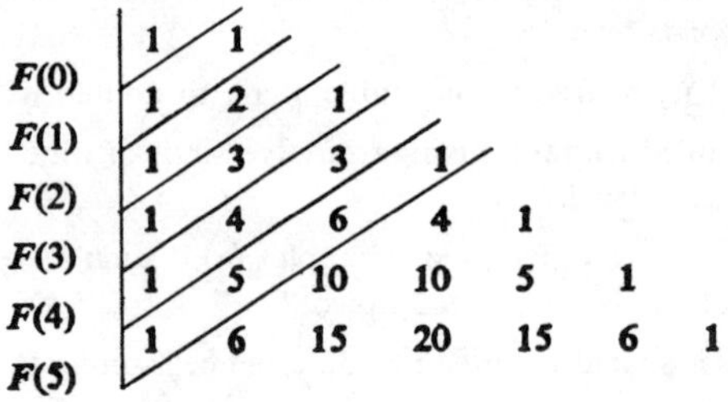

$$F(0) = 1$$

$$F(1) = 2$$

$$F(2) = \binom{3}{0} + \binom{2}{1}$$

$$F(3) = \binom{4}{0} + \binom{3}{1} + \binom{2}{2}$$

$$F(4) = \binom{5}{0} + \binom{4}{1} + \binom{3}{2}$$

$$F(5) = \binom{6}{0} + \binom{5}{1} + \binom{4}{2} + \binom{3}{3}$$

$$F(n) = \binom{n+1}{0} + \binom{n}{1} + \binom{n-1}{2} + \cdots + a_n; \tag{6}$$

$$\text{if } n \text{ is even } a_n = \binom{n/2+1}{n/2} \text{ and } a_n = \binom{\frac{n+1}{2}}{\frac{n+1}{2}} \text{ if } n \text{ is odd.}$$

Show that

$$F(n) = F(n-1) + F(n-2) \tag{7}$$

by substituting (6) in the right side of (7) and by using the identity (3).

17 Suppose there are n symbols, s, in a row and that each s can be a plus sign or a minus sign. There are 2^n cases in all. Let $F(n)$ denote the number of cases in which no two minus signs are adjacent.

a) Verify the following statements:

1) Of the $F(n)$ sequences described above there are $F(n-1)$ that begin with a plus sign and there are $F(n-2)$ that begin with a minus sign.

2) $F(n)$ satisfies (7) of exercise 16.

b) Show that $n-r$ plus signs and r minus signs can be arranged in a row so that no two minus signs are adjacent in $\binom{n-r+1}{r}$ ways. By considering successively the types: n plus signs, $n-1$ plus signs and 1 minus sign, $n-2$ plus signs and 2 minus signs etc. show that (1) of exercise 16 is valid.

18 How many different paths of length 15 units are there from one intersection point in a three-dimensional rectangular steel framework to another which is located four units to the right, five units back and six units up?

19 a) With reference to example 3 how many keys must the president have in order that he and any general can open the safe?

b) Suppose that a specified general G_5 duplicates a key that he did not already have. How many sets of three (including G_5) can open the safe?

c) Rework example 3 under the assumption that the safe can be opened by a majority or any set of three which includes a specified general G, if G cannot open the safe alone.

20 Show that there are $\binom{n+5}{5}$ distinguishable results of a throw with n indistinguishable dice.

21 How many different combinations of twelve poker chips are there if each chip is colored red, white, or blue?

5 Solutions of linear equations

Consider the solutions of the equation

$$x_1 + x_2 + \cdots + x_{10} = 6$$

in nonnegative integers. In counting solutions we shall say that

$$x_1 = 1\ x_2 = 1\ x_3 = 1\ x_4 = 1\ x_5 = 1\ x_6 = 0\ x_7 = 1\ x_8 = 0\ x_9 = 0\ x_{10} = 0$$

$$x_1 = 0\ x_2 = 1\ x_3 = 1\ x_4 = 1\ x_5 = 1\ x_6 = 0\ x_7 = 1\ x_8 = 1\ x_9 = 0\ x_{10} = 0$$

are distinct solutions because as ordered 10-tuples they are distinct. Hence the number of distinct solutions is the same as the number of ways of placing 6 identical balls in 10 different boxes if we interpret $x_i = k$ to mean that box i contains k balls. From example 4.8 the number of solutions is just $\binom{15}{6}$.

More generally, let us find the number of distinct solutions of the equation

$$x_1 + x_2 + \cdots + x_k = n \tag{1}$$

in nonnegative integers. Under the above interpretation this is just the number of ways of placing n similar balls into k different boxes. Alternatively, the number of distinct solutions of (1) in nonnegative integers is the number of n-combinations with repeats permitted of k distinct things when x_i is the number of times the ith object is included in the combination. By the same procedure as used in example 4.8 (or using theorem 4.2) the answer is

$$\binom{n+k-1}{n}. \tag{2}$$

The number of distinct solutions of (1) in positive integers is

$$\binom{n-1}{k-1} \tag{3}$$

by example 4.9 for $n \geqslant k$. This problem is equivalent to finding the number of ways of placing n similar balls in k different boxes so that no box remains empty.

We now turn to the problem of finding the number of distinct solutions of (1) subject to certain restrictions. For example, how many different solutions in integers greater than 4 are there of the equation

$$x + y + z = 25? \tag{4}$$

Since each variable must be greater than 4, subtracting 4 from each variable gives a new set of variables which must be positive. Let

$$r = x - 4, \quad s = y - 4, \quad t = z - 4. \tag{5}$$

For example, an allowable solution of (4) is $x = 6$, $y = 5$, $z = 14$. r, s, t, then become the positive numbers 2, 1, 10 respectively. Conversely supposing that r, s, t are defined by (5) then x, y, z are uniquely determined by

$$x = r + 4, \quad y = s + 4, \quad z = t + 4. \tag{6}$$

Substituting (6) into (4) yields

$$r + s + t = 13. \tag{7}$$

Hence equations (4) and (7) are related by the transformations (6) (or (5)). Each solution of (4) uniquely corresponds to a solution of (7) and conversely. For example, $x = 6, y = 5, z = 15$ corresponds to $r = 2, s = 1, t = 10$ and $r = 3, s = 4, t = 6$ corresponds to $x = 7, y = 8, z = 10$. Thus corresponding to each solution of (4) in integers greater than 4 there is a solution of (7) in positive integers and conversely. Therefore the number of distinct solutions of (4) in integers greater than 4 is the same as the number of distinct solutions of (7) in positive integers. By (3) this number is $\binom{12}{2}$.

This idea still works if each variable is subject to a different condition. For example, let us consider the problem: How many distinct solutions are there of the equation

$$x + y + z + w = 58 \tag{8}$$

in integers satisfying

$$x > -5, y > 6, z > -15, w > 9? \tag{9}$$

The required transformation is

$$r = x + 5, \quad s = y - 6, \quad t = z + 15, \quad u = w - 9 \tag{10}$$

or

$$x = r - 5, \quad y = s + 6, \quad z = t - 15, \quad w = u + 9. \tag{11}$$

Substituting (11) into (8) yields

$$r + s + t + u = 63. \tag{12}$$

Any solution of (8) in integers subject to the restrictions (9) corresponds uniquely to a solution in positive integers of (12) and conversely (by using (10) or (11)). For example, the solution $x = 5$, $y = 10$, $z = 13$, $w = 30$ corresponds to the solution $r = 10$, $s = 4$, $t = 28$, $u = 21$. Therefore the number of distinct solutions of (8) subject to the restrictions (9) equals the number of distinct solutions of (12) in positive integers. By (3) this number is $\binom{62}{3}$.

Suppose that the number of distinct solutions of (1) in integers subject to the conditions

$$x_1 > c_1, \quad x_2 > c_2, \ldots, x_k > c_k \tag{13}$$

are required, where $c_1, c_2, \ldots, c_k$ are k fixed integers. The proper transformation is

$$y_1 = x_1 - c_1, \quad y_2 - c_2, \ldots, y_k = x_k - c_k \tag{14}$$

or

$$x_1 = y_1 + c_1, \quad x_2 = y_2 + c_2, \ldots, x_k = y_k + c_k. \tag{15}$$

Substituting (15) into (1) yields

$$y_1 + y_2 + \cdots + y_k = n - c_1 - c_2 - \cdots - c_k. \tag{16}$$

Hence the number of distinct solutions in integers of (1) subject to the restrictions (13) is equal to the number of distinct solutions of (16) in positive integers. Another application of (3) yields the answer

$$\binom{n - c_1 - c_2 - \cdots - c_k - 1}{k - 1}.$$

We will return to this subject in Chapter III.

Exercises

1 How many distinct solutions are there of $x + y + z + w = 61$ (a) in nonnegative integers? (b) in positive integers?

2 Prove that the number of distinct solutions of (1) in positive integers equals the number of distinct solutions in positive integers of the equation $x_1 + x_2 + \cdots + x_{n+k-1} = n$.

3 Find the number of distinct triples (x, y, z) of nonnegative integers satisfying the inequality $x + y + z < 25$.

4 How many integers between 1 and 1 000 inclusive have a sum of digits (a) equal to 5? (b) less than 5?

5 Write out the one-to-one correspondence between the solutions of $x + y + z = 15$ in integers greater than 3, and the solutions of $r + s + t = 6$ in positive integers. In other words for each solution (x, y, z) find the unique corresponding solution (r, s, t).

6 Find the number of solutions of $x_1 + x_2 + \cdots + x_6 = 75$ in positive integers (a) with $x_1 > 20$. (b) with $x_1 > 10$ and $x_2 > 6$.

7 Find the number of solutions of $x + y + z + w = 57$ subject to $x > -6$, $y > 7$, $w > -15$, $z > 10$.

8 Show that the number of combinations of $1, 2, \ldots, n$ taken r at a time without including repeats or consecutive numbers is just the number of distinct solutions in integers of $n = x_1 + \cdots + x_{r+1}$ where $x_1 \geqslant 1$, $x_2 \geqslant 2$, $x_3 \geqslant 2, \cdots, x_r \geqslant 2$, $x_{r+1} \geqslant 0$?

6 Division into classes

Whenever we choose r objects from n there are $n - r$ objects left over. Hence the number of ways of choosing r objects from n distinct objects is equal to the number of ways of dividing n objects into two classes containing $k_1 = n$ and $k_2 = n - r$ objects. Therefore the number of ways n distinct things can be divided into two classes of k_1 and k_2 respectively is $\dfrac{n!}{k_1!\,k_2!}$ according to theorem 4.1. The next theorem generalizes this result.

THEOREM 1 *The number of ways that n distinct things can be divided into r* distinguishable *classes containing $k_1, k_2, \ldots, k_r$ objects respectively, where $k_1 + k_2 + \cdots + k_r = n$, is*

$$\frac{n!}{k_1!\,k_2!\,k_3!\cdots k_r!}. \tag{1}$$

Proof The objects in the first group can be selected in $\binom{n}{k_1}$ ways. From the remaining $n - k_1$ things, the members of the second group can be chosen in $\binom{n-k_1}{k_2}$ ways Then the objects in the third group can be chosen in

$\binom{n-k_1-k_2}{k_3}$ ways since there are $n - k_1 - k_2$ remaining objects and so on. Finally the objects in the last group can be chosen in $\binom{n-k_1-k_2\cdots-k_{r-1}}{k_r}$ ways. By the corollary to theorem 2.1 we see that the number of ways is

$$\binom{n}{k_1}\binom{n-k_1}{k_2}\binom{n-k_1-k_2}{k_3}\cdots\binom{n-k_1-k_2-\cdots-k_{r-1}}{k_r}$$

$$= \left(\frac{n!}{k_1!\,(n-k_1)!}\right)\left(\frac{(n-k_1)!}{k_2!\,(n-k_1-k_2)!}\right)\left(\frac{(n-k_1-k_2)!}{k_3!\,(n-k_1-k_2-k_3)!}\right)\cdots$$

$$\left(\frac{(n-k_1-k_2\cdots-k_{r-1})!}{k_r!\,(n-k_1-k_2\cdots-k_{r-1}-k_r)!}\right) = \frac{n!}{k_1!\,k_2!\,k_3!\cdots k_r!}$$

If all the k_i $(i = 1, \ldots, r)$ are different the classes will be different since they contain different numbers of elements. However *if* $k_1 = k_2 = \cdots = k_r$ the classes need not be different and we must postulate this for (1) to be valid. If the classes are similar we must divide (1) by $r!$ since the ways of arranging the r classes must be ignored. To illustrate this point let us divide the numbers 1, 2, 3, 4 into 2 groups of 2 each. Let (1, 2), (3, 4) denote that the numbers 1 and 2 are in the first group while 3 and 4 are in the second group. If we assume the groups are distinguishable we obtain $\frac{4!}{2!\,2!} = 6$ possibilities, namely:

(1, 2), (3, 4); (3, 4), (1, 2); (1, 3), (2, 4); (2, 4), (1, 3); (2, 3), (1, 4); (1, 4), (2, 3).

However, if the groups are indistinguishable (1, 2), (3, 4) is the same as (3, 4), (1, 2) etc. Therefore there are only 3 ways of dividing the objects into groups, namely:

(1, 2), (3, 4); (1, 3), (2, 4); (2, 3), (1, 4).

For contrast, suppose the above numbers are divided into 2 distinguishable groups the first containing 3 numbers and the other 1 number. The $\frac{4!}{3!\,1!} = 4$ possibilities are:

(1, 2, 3), (4); (1, 2, 4), (3); (1, 3, 4), (2,); (2, 3, 4), (1).

As another example, 10 men can be divided into 2 teams of five each to play against each other in $\frac{10!}{2!\,5!\,5!}$ ways. The following example also illustrates this point.

Example 1 (a) How many different hands can be held by 4 men playing bridge (13 cards each)? (b) In how many ways can a pack of 52 cards be arranged in 4 heaps of 13 cards each?

In (a) the hands must be placed on the table in a definite order. The number of ways is $\frac{52!}{13!\,13!\,13!\,13!}$. In (b) no regard is paid to the order of the heaps and so the number of ways is $\frac{52!}{(13!)^4\,4!}$.

The alert reader will have observed that (1) is the same as equation 3.5. This is not a mere coincidence. Theorem 1 can be derived by using theorem 3.2. Write symbols for the n objects in a straight line (see below). Now consider a particular division in r groups. If an object belongs to the i-th group write the number i underneath it. For example, suppose there are 3 groups G_1, G_2, G_3 and seven objects, denoted by a, b, c, d, e, f, g. The following table illustrates this code.

$a\ b\ c\ d\ e\ f\ g$	G_1	G_2	G_3
1 1 2 2 2 3 3	a, b	c, d, e	f, g
1 1 1 1 1 1 1	a, b, c, d, e, f, g		
3 3 3 3 3 3 3			a, b, c, d, e, f, g
3 2 1 3 2 1 1	c, f, g	b, e	a, d

The 1's under a and b in the first line indicate that a and b belong to G_1; the 2's under c, d, e show that these elements belong to G_2 while the remainder, marked with a 3, belong to G_3. It is evident that every arrangement of k_1 ones, k_2 twos, ..., k_r r's corresponds uniquely to a division into groups and vice versa. The number of different ways of arranging k_1 objects of the first kind k_2 objects of the second kind etc., is given by theorem 3.2 which yields (1).

Example 2 Consider the placement of 10 distinguishable balls into 7 different cells. Find the total number of distributions of balls in cells such that the occupancy numbers coincide with 3, 3, 2, 1, 1, 0, 0 in some order, that is,

3 balls in a cell, 3 balls in another cell, 2 balls in a third cell etc. Repeat for occupancy numbers 6, 1, 1, 1, 1, 0, 0 and 6, 4, 0, 0, 0, 0, 0.

Each permutation $(i_1, i_2, \ldots, i_7)$ of the occupancy numbers corresponds to a particular way of distributing the balls into the cells by assigning i_k balls to cell k ($k = 1, 2, \ldots, 7$). For instance, (3, 3, 2, 1, 1, 0, 0) denotes that cell 1, 2, 3, 4, 5, 6, 7 contains 3, 3, 2, 1, 1, 0, 0 balls respectively. The total number of different ways that the cells can be occupied is $\dfrac{7!}{2!\,2!\,2!}$ by theorem 3.2. Given that the cells are occupied in a particular order we must now specify which balls go in which cells. For example, if the cells are occupied as follows (3, 3, 2, 1, 1, 0, 0) we must divide the 7 balls into 7 groups containing 3, 3, 2, 1, 1, 0, 0 balls respectively. This can be done in $\dfrac{10!}{3!\,3!\,2!\,1!\,1!\,0!\,0!}$ ways according to theorem 1. Note that this division into groups is independent of the permutation which is used. Hence the number of distributions of balls in cells is $\left(\dfrac{7!}{2!\,2!\,2!}\right)\left(\dfrac{10!}{3!\,3!\,2!}\right)$.

The remaining parts are treated similarly; the answers are $\left(\dfrac{7!}{4!\,2!}\right)\left(\dfrac{10!}{6!}\right)$ and $\left(\dfrac{7!}{5!}\right)\left(\dfrac{10!}{6!\,4!}\right)$ respectively. The first factor in each of the products is the number of permutations of the occupancy numbers while the second factor represents the number of ways the balls can be divided into groups corresponding to a particular permutation.

Exercises

1 In how many ways can 12 people be divided into 6 pairs?

2 In how many ways can 6 different toys be distributed to A, B, C so that A has 3, B has 2 and C has 1?

3 8 men are to play 4 singles at tennis, the games being simultaneous. In how many ways can the games be arranged?

4 In how many ways can 3 boys divide 12 pears, each taking 4 if

a) the pears are all of the same size?

b) the pears are of different sizes?

5 Suppose an elevator starts with seven passengers and stops at ten floors. The various arrangements of unloading may be denoted by symbols like (3, 2, 2), to be interpreted as the event that 3 passengers leave together at a certain floor, two other passengers at another floor and the last two at still another floor. Find the number of ways the 15 possible arrangements ranging from (7) to (1, 1, 1, 1, 1, 1, 1) can occur.

7 Selections taking any number at a time

We now find the number of ways of selecting any number of objects (from 0 to n) from n given objects. First we assume that the objects are distinct.

Example 1 What is the greatest number of different amounts that can be made up by a selection from 3 given weights?

Write the 3 weights, W_1, W_2, W_3 in a row with a 1 underneath the weight if it is in a particular selection and a 0 underneath otherwise. For example,

W_1	W_2	W_3	Weights selected
1	1	0	W_1 W_2
1	1	1	W_1 W_2 W_3
0	1	0	W_2

Each selection of weights corresponds uniquely to a 3-digit number where each digit is 1 or 0 and conversely. The number of such 3-digit numbers is $2 \times 2 \times 2 = 8$ by theorem 3.3. This is just another way of saying that for each object we have 2 choices; either we select it or we reject it. Note that if W_1, W_2, W_3 is 1, 2, 3 respectively the maximum number would not be achieved since $W_1 + W_2 = W_3$. The maximum number is realized by using the weights 1, 2, and 4.

THEOREM 1 *The number of selections from n unlike things, taking any number at a time, is* 2^n.

Proof We have two choices for each object; either we reject it or select it. Hence n objects can be disposed of in

$$2 \times 2 \times \cdots (n \text{ factors}) = 2^n$$

ways.

Note that the number of selections, if at least 1 thing is chosen, is $2^n - 1$ since the above selections include the situation in which no things are chosen.

We can calculate the number of selections of any size in a different manner and thus derive an interesting algebraic identity. Each selection contains 0 or 1 or 2 $\cdots$ or n objects. The number of ways of selecting k objects from n is $\binom{n}{k}$. Therefore the total number of selections taking any number at a time is

$$\binom{n}{0} + \binom{n}{1} + \binom{n}{2} + \cdots + \binom{n}{k} + \cdots + \binom{n}{n} = 2^n. \tag{1}$$

This result will be derived analytically in the following chapter.

We consider the situation in which the objects are not distinct; for instance, selections from 1, 1, 2, 3, 3. If we choose the first 1 and 2 we have the pair 1, 2. This is considered to be the same as choosing the second 1 and 2 which is just 1, 2 again. Distinct selections appear in the diagram below.

Selection of ones	Selection of twos	Selection of threes	Final outcome
none	none	none	none
		3	3
		3, 3	3, 3
	2	none	2
		3	2, 3
		3, 3	2, 3, 3
1	none	none	1
		3	1, 3
		3, 3	1, 3, 3
	2	none	1, 2
		3	1, 2, 3
		3, 3	1, 2, 3, 3
1, 1	none	none	1, 1
		3	1, 1, 3
		3, 3	1, 1, 3, 3
	2	none	1, 1, 2
		3	1, 1, 2, 3
		3, 3	1, 1, 2, 3, 3

THEOREM 2 *Given p_1 similar things of 1 kind, p_2 similar things of another kind, …, and p_r similar things of an r-th kind, the number of selections which can be made, taking any number at a time, is*

$$(p_1 + 1)(p_2 + 1)(p_3 + 1)\cdots(p_r + 1). \tag{2}$$

Proof The p_1 similar things can be disposed of in $p_1 + 1$ ways since we can select either 0 or 1 or 2 or $\cdots$ or p_1 of them. Similarly the other groups can be disposed of in $(p_2 + 1)$ ways, $(p_3 + 1)$ ways etc. Therefore the number of ways of disposing of them all is given by (2) according to the corollary to theorem 2.1.

Note that this includes the case in which all are rejected. Therefore the number of selections, taking at least 1 thing, is

$$(p_1 + 1)(p_2 + 1)(p_3 + 1)\cdots(p_r + 1) - 1.$$

Example 2 In how many ways can 200 copies of one book, 250 of another, 300 of a third, and 200 of a fourth be distributed to 2 book stores?

We assume that either store can take any number of books, but neither can take all. Therefore, by the above theorem, the number of ways is (201) (251) (301) (201) − 2.

Exercises

1 In how many ways can any number of letters be selected from the letters *a, a, a, a, a, b, b, b, b, c, c, c, d, d, e, f, g*?

2 How many selections can be made from 10 different books, if any number of them may be taken?

3 How many distinct positive divisors has the number 155, 925?

4 What is the greatest number of different amounts that can be weighed with five weights, when each weight may be put in either scale? Give an example of weights which will realize this maximum.

5 In how many ways can 144 be expressed as the product of 2 positive integers?

Miscellaneous problems

1 Verify the following assertions:

a) If 5 balls are placed in 4 boxes, at least one box contains 2 or more balls; more generally, if $n + 1$ balls are placed in n boxes, at least one of the boxes contains 2 or more balls.

b) If k_{n+1} balls are placed in n boxes, at least one of the boxes contains $k + 1$ or more balls.

2 A certain city has a population of 2,000,000. Given that no human being has more than 250,000 hairs on his head, use problem 1 to find the largest integer n that can be used in the following assertion. There are n people in the city with the same number of hairs on their heads.

3 Assume that at least one of a_i and b_i has a certain property P $(i = 1, 2, \dots, 5)$. Show that at least three of a_1, a_2, a_3, a_4, a_5 or at least three of b_1, b_2, b_3, b_4, b_5 have property P.

4 Suppose there are 6 points in a plane, no three of which lie on a straight line. The points are joined in pairs by line segments colored either red or white. Prove that there are three points such that the arcs of the triangle they form are of the same color. Such a triangle is called a *chromatic* triangle. (Hint: choose a point P_0 and consider the five arcs from P_0 to the remaining five points. Three of these arcs must be the same color (say white); let these be denoted by P_0P_1, P_0P_2, P_0P_3. One of the arcs P_1P_2, P_1P_3, or P_2P_3 must be white.)*†

5 Find an example of a set of five points, with red and white arcs joining the pairs so that there is no chromatic triangle.

6 Show that among any 6 persons it is possible to find three who are acquainted or it is possible to find three no two of whom are acquainted.

7 Consider seventeen points in the plane, no three of which lie on a straight line. Each line segment joining a pair of points is colored red, white, or blue. Prove that there is a chromatic triangle.†

---

* F. P. Ramsey, "On a problem of formal logic", *Proc. Lond. Math. Soc.*, 2nd series, 30 (1931).

† R. E. Greenwood and A. M. Gleason, "Combinatorial relations and chromatic graphs", *Can. J. of Math.*, 7 (1955).

CHAPTER II

The multinomial theorem

1 Summation and product notation

IN THE LAST chapter, it was often necessary to write out the sums and products of a finite sequence of numbers. For example, ${}_nP_r = n(n-1)(n-2) \cdots (n-r+1)$ and one of the hypotheses of theorem 1.3.2 states that $k_1 + k_2 + \cdots + k_r = n$. Instead of writing out each term it is convenient to have a shorthand notation

We denote the finite series $a_1 + a_2 + \cdots + a_n$ by

$$\sum_{r=1}^{n} a_r \quad \text{or} \quad \sum_{1}^{n} a_r.$$

The symbol $\sum$ is the Greek letter "sigma" and indicates that the sequence of numbers $a_1, a_2, \ldots, a_n$ is to be summed. 1 and n are the lower and upper limits of the series. These limits, of course, can be any finite number. For example,

$$\sum_{r=-2}^{m} a_r = a_{-2} + a_{-1} + a_0 + a_1 + \cdots + a_m,$$

$$\sum_{r=3}^{10} a_r = a_3 + a_4 + a_5 + \cdots + a_{10},$$

$$\sum_{r=e}^{m} a_r = a_e + a_{e+1} + \cdots + a_m.$$

In the last example, the sequence $a_e, a_{e+1}, a_{e+2}, \ldots, a_m$ to be summed is obtained from the term a_r by just setting r equal to the lower limit e, then to $e+1, e+2, e+3, \ldots$ successively until the upper limit m is reached. r is called a dummy variable since we could also denote $a_1 + a_2 + \cdots + a_n$ by $\sum_{i=1}^{n} a_i$ or $\sum_{k=1}^{n} a_k$ etc. If these last two sums are written out in the long form neither the dummy variable i nor k appears.

Example 1

a) $k_1 + k_2 + \cdots + k_r = \sum_{i=1}^{r} k_i.$

b) $4 + 5 + 6 + \cdots + 10 = \sum_{r=4}^{10} r.$

c) $7^2 + 8^2 + \cdots + m^2 = \sum_{r=7}^{m} r^2.$

d) $\sum_{r=1}^{n} c = nc \quad (a_r = c \text{ for all } r).$

e) $\sum_{-2}^{m} 3^r = 1/3^2 + 1/3 + 1 + 3 + 9 + \cdots + 3^m.$

f) $\sum_{r=1}^{n} (3/10)^r = 3/10 + (3/10)^2 + \cdots + (3/10)^n.$

g) $\sum_{r=1}^{m} a_r = a_1 + \sum_{r=2}^{m} a_r = \sum_{k=1}^{n} a_k + \sum_{k=n+1}^{m} a_k \quad (\text{where } 1 \leqslant n < m).$

Every symbol of the form $\sum_{i=1}^{n} a_i$ may be quickly written in the form $a_1 + a_2 + \cdots + a_n$. It is not easy to reverse this process. For example, it is no simple matter to put the series

$$1 + 3 - 2 - 6 + 78 - 43 + 193$$

in the form $\sum_{r=1}^{n} a_r$. In order to use the abbreviated form it is necessary to express the rth term of the series as a simple function of r $(1 \leqslant r \leqslant n)$. The notation a_r is used instead of the usual functional notation $a(r)$. Finding the general term a_r of the series $a_1 + a_2 + \cdots + a_n$ is equivalent to finding the general term a_r of the associated sequence $a_1, a_2, \ldots, a_n$.

Example 2

(a) 1, 2, 3, 4, 5 has as its general term $a_r = r$ since $a_1 = 1, a_2 = 2, a_3 = 3, a_4 = 4$, and $a_5 = 5$.

(b) 1, 3, 5, 7, 9, ..., 31 has as its general term $a_r = 2r - 1$ since $a_1 = 1, a_2 = 3, a_3 = 5$ etc.

(c) $a_r = r^2$ is the general term of the sequence $1^2, 2^2, 3^2, \ldots, 101^2$.
(d) $a_r = 2$ is the general term of the sequence 2, 2, 2, ..., 2.
(e) $a_r = (-1)^r r + r$ is the general term of the sequence 0, 4, 0, 8, 0, 12, 0.

Once the general term is known it is easy to write down the abbreviated form. For example,

$$1 + 2 + 3 + 4 + 5 = \sum_{r=1}^{5} r;$$

$$1 + 3 + 5 + \cdots + 31 = \sum_{r=1}^{16} (2r - 1).$$

A series which appears quite often is the *arithmetic progression*, $a + (a + d) + (a + 2d) + \cdots$. The general term is $a_r = a + (r - 1)\,d$ where d is the difference between any two successive terms, that is, $a_r - a_{r-1} = d$. We will find an explicit expression for the sum s of n terms of this series, that is,

$$s = a + [a + d] + [a + 2d] + \cdots + [a + (n - 1)\,d]. \tag{1}$$

To do this we rewrite the series in the reverse order—namely,

$$s = [a + (n - 1)\,d] + [a + (n - 2)\,d] + [a + (n - 3)\,d] + \cdots + a. \tag{2}$$

Adding (1) and (2) term by term yields

$$2s = [2a + (n - 1)\,d] + [2a + (n - 1)\,d] + [2a + (n - 1)\,d] + \cdots$$
$$+ [2a + (n - 1)\,d] = n\,[2a + (n - 1)\,d].$$

Solving for s,

$$s = \frac{n}{2}\,[2a + (n - 1)\,d]. \tag{3}$$

In particular,

$$\sum_{1}^{n} r = 1 + 2 + 3 + \cdots + n = \frac{n\,(n + 1)}{2} \tag{4}$$

as follows from (3) with $a = d = 1$.

Another useful series is the geometric progression. In this series, $a_r = ar^{n-1}$. We will find an explicit expression for the sum

$$s = a + ar + ar^2 + \cdots + ar^{n-1} \tag{5}$$

of n terms of this series. Multiplying (5) by r yields

$$rs = ar + ar^2 + \cdots + ar^{n-1} + ar^n. \tag{6}$$

Subtracting (6) from (5) yields

$$(1 - r)\, s = a - ar^n \quad \text{or} \quad s = \frac{a\,(1 - r^n)}{1 - r}.$$

To denote the product

$$P_n = a_1 \times a_2 \times \cdots \times a_n$$

of n terms we write

$$P_n = \prod_{r=1}^{n} a_r \quad \text{or} \quad P_n = \prod_{1}^{n} a_r.$$

The symbol $\prod$ is the Greek letter "pi" and indicates that the sequence whose general term is a_r is to be multiplied together. The same discussion holds for finite products as for finite sums if we replace $\sum$ by $\prod$ and references to addition by references to multiplication.

Example 3

a) $\prod_{i=1}^{n} i = \prod_{r=1}^{n} r = \prod_{k=1}^{n} k = n!.$

b) $\prod_{i=1}^{n} k^2 = (n!)^2.$

c) $\prod_{i=1}^{n} \frac{i+1}{i} = \left(\frac{2}{1}\right)\left(\frac{3}{2}\right)\left(\frac{4}{3}\right)\cdots\left(\frac{n+1}{n}\right) = n + 1.$

d) $\prod_{i=1}^{n} c = c^n.$

Certain formal rules are often helpful in manipulating sums and products.

THEOREM 1

a) $\sum_{r=1}^{n} (a_r + b_r) = \sum_{1}^{n} a_r + \sum_{1}^{n} b_r.$

b) $\sum_{r=1}^{n} (ca_r) = c\left(\sum_{r=1}^{n} a_r\right)$, *where c is a constant.*

Proof

a) $$\sum_{r=1}^{n} (a_r + b_r) = (a_1 + b_1) + (a_2 + b_2) + (a_3 + b_3) + \cdots + (a_n + b_n)$$
$$= (a_1 + a_2 + \cdots + a_n) + (b_1 + b_2 + \cdots + b_n)$$
$$= \sum_{r=1}^{n} a_r + \sum_{r=1}^{n} b_r.$$

b) $$\sum_{r=1}^{n} (ca_r) = ca_1 + ca_2 + ca_3 + \cdots + ca_n = c\,(a_1 + a_2 + \cdots + a_n)$$
$$= c\left(\sum_{r=1}^{n} a_r\right).$$

THEOREM 2

a) $$\prod_{i=1}^{n} a_i b_i = \left(\prod_{i=1}^{n} a_i\right)\left(\prod_{i=1}^{n} b_i\right).$$

b) $$\prod_{i=1}^{n} ca_i = c^n\left(\prod_{i=1}^{n} a_i\right).$$

The proofs follow immediately by expanding both sides of the above equations.

Example 4

1 Find an analytic formula for $\sum_{1}^{n} r^2$.
Consider the identity

$$(r + 1)^3 - r^3 = 3r^2 + 3r + 1.$$

Summing both sides of the above identity yields

$$\sum_{1}^{n} [(r + 1)^3 - r^3] = \sum_{1}^{n} (3r^2 + 3r + 1) = 3\sum_{1}^{n} r^2 + 3\sum_{1}^{n} r + n. \tag{7}$$

Since

$$\sum_{1}^{n} [(r + 1)^3 - r^3] = [(n + 1)^3 - n^3] + [n^3 - (n - 1)^3]$$
$$+ [(n - 1)^3 - (n - 2)^3] + \cdots + [2^3 - 1^3]$$
$$= (n + 1)^3 - 1 = n^3 + 3n^2 + 3n$$

and $3\sum_1^n r = \dfrac{3n(n+1)}{2}$, (7) yields

$$3\sum_1^n r^2 = n^3 + 3n^2 + 3n - n - \frac{3n(n+1)}{2}$$

$$= \frac{1}{2}[2n^3 + 6n^2 + 6n - 2n - 3n^2 - 3n]$$

$$= \frac{1}{2}[2n^3 + n + 3n^2] = \frac{n(n+1)(2n+1)}{2}.$$

Therefore,

$$\sum_1^n r^2 = \frac{n(n+1)(2n+1)}{6}.$$

2 By a similar procedure, using the identity

$$(r+1)^4 - r^4 = 4r^3 + 6r^2 + 4r + 1$$

it can be shown that

$$\sum_1^n r^3 = \left[\frac{n(n+1)}{2}\right]^2.$$

Double and multiple sums and products have a similar interpretation.

Example 5

a) $$\sum_{k=1}^{4}\sum_{j=0}^{3} a_{jk} = \sum_{k=1}^{4}(a_{0k} + a_{1k} + a_{2k} + a_{3k})$$

$$= (a_{01} + a_{11} + a_{21} + a_{31}) + (a_{02} + a_{12} + a_{22} + a_{32})$$
$$+ (a_{03} + a_{13} + a_{23} + a_{33}) + (a_{04} + a_{14} + a_{24} + a_{34})$$
$$= (a_{01} + a_{02} + a_{03} + a_{04}) + (a_{11} + a_{12} + a_{13} + a_{14})$$
$$+ (a_{21} + a_{22} + a_{23} + a_{24}) + (a_{31} + a_{32} + a_{33} + a_{34})$$
$$= \sum_{j=0}^{3}(a_{j1} + a_{j2} + a_{j3} + a_{j4}) = \sum_{j=0}^{3}\sum_{k=1}^{4} a_{jk}.$$

b) $$\sum_{j=2}^{3}\sum_{k=0}^{1}\sum_{m=1}^{2} a_{jkm} = \sum_{j=2}^{3}\sum_{k=0}^{1}(a_{jk1} + a_{jk2}) = a_{201} + a_{301} + a_{211}$$
$$+ a_{311} + a_{202} + a_{302} + a_{212} + a_{312}.$$

c) $\prod_{i=1}^{2} \prod_{k=0}^{3} (i + k) = \prod_{i=1}^{2} (i)(i + 1)(i + 2)(i + 3)$

$$= 1 \cdot 2^2 \cdot 3^2 \cdot 4^2 \cdot 5 = \prod_{k=0}^{3} \prod_{i=1}^{2} (i + k).$$

THEOREM 3

a) $\sum_{j=1}^{m} \sum_{k=1}^{n} a_{jk} = \sum_{k=1}^{n} \sum_{j=1}^{m} a_{jk}.$

b) $\prod_{j=1}^{m} \prod_{k=1}^{n} a_{jk} = \prod_{k=1}^{n} \prod_{j=1}^{m} a_{jk}.$

The proof follows immediately upon expanding both sides of the equation and comparing terms.

Transformations of the dummy variable r of the form $k = r + b$, with b an integer and k as the new dummy variable, are useful in manipulation of sums.

Example 6

(a) To evaluate $\sum_{r=2}^{203} (r - 2)$, set $r - 2 = k$. We let $r - 2 = k$ in both the limits and the expression being summed. When $r = 2$ the new dummy variable $k = 2 - 2 = 0$ and so the new lower limit is 0. Similarly the new upper limit becomes $k = 203 - 2 = 201$. The general term $a_r = r - 2$ is transformed into $a_{k+2} = (k + 2) - 2 = k$. Thus the sum transforms into $\sum_{k=0}^{201} k = (101)(201)$. We leave it to the reader to verify that the value of these two sums are the same.

(b) Find $\sum_{i=5}^{m+5} i + \sum_{j=-2}^{m-2} (j + 1)$. Making the transformations $i - 5 = r$ and $j + 2 = s$, we obtain

$$\sum_{r=0}^{m} (r + 5) + \sum_{s=0}^{m} (s - 1).$$

If we replace the dummy variable r by s in the first sum and then apply theorem 1, we have

$$\sum_{s=0}^{m} (s + 5) + \sum_{s=0}^{m} (s - 1) = \sum_{s=0}^{m} (s + 5 + s - 1) = \sum_{s=0}^{m} (2s + 4)$$

$$= 2 \sum_{s=0}^{m} s + 4m = m(m + 1) + 4m = m^2 + 5m.$$

The reader, perhaps without being aware of it, is already familiar with series which do not have a finite number of terms. Consider $1/3 = 0.3333\cdots$.

This may be written as

$$3/10 + 3/100 + 3/1000 + \cdots + .$$

The dots indicate that the series never ends (is infinite). Often, in order to use this series in calculations we "round off" by writing

$$1/3 = 0.3333 = 3/10 + 3/100 + 3/1000.$$

If greater accuracy is required, we might write

$$1/3 = 0.33333 = 3/10 + 3/100 + \cdots + 3/10^5.$$

We realize, of course, that neither of these two finite series really equals 1/3 but only approximates it.

A short-hand symbol for infinite series is used which is similar to that used for finite series. We denote the infinite series

$$a_1 + a_2 + \cdots + \quad \text{by} \quad \sum_{i=1}^{\infty} a_i \quad \text{or} \quad \sum_{1}^{\infty} a_i.$$

The first n terms of the infinite series $\sum_{i=1}^{\infty} a_i$ is the finite series $s_n = \sum_{i=1}^{n} a_i$. s_n is called the *nth partial sum of the series* or the *sum to n terms of the series.*

A procedure similar to that used for finite series is used to write an infinite series in the shorthand form. To write 0.333... in this form we observe that the general term of the associated sequence $3/10, 3/10^2, 3/10^3, \ldots,$ is $3/10^n$ and so

$$1/3 = \sum_{r=1}^{\infty} 3/10^n.$$

Similar remarks hold for infinite products.

Example 7

1 $\sum_{3}^{\infty} (\tfrac{1}{2})^n = (\tfrac{1}{2})^3 + (\tfrac{1}{2})^4 + (\tfrac{1}{2})^5 + \cdots.$

2 $\sum_{-4}^{\infty} a_i = a_{-4} + a_{-3} + a_{-2} + a_{-1} + a_0 + a_1 + a_2 + a_3 + \cdots.$

3 $\sum_{-\infty}^{\infty} b_i = \cdots + b_{-3} + b_{-2} + b_{-1} + b_0 + b_1 + b_2 + b_3 + \cdots.$

4 $\prod_{i=2}^{\infty} a_i = a_2 a_3 a_4 \ldots.$

5 $\prod_{n=1}^{\infty} \dfrac{n}{n+1} = \dfrac{1}{2} \cdot \dfrac{2}{3} \cdot \dfrac{3}{4} \cdot \dfrac{4}{3} \cdots$

The reader should be wary of applying theorems concerning finite series (products) to infinite series (products). With every finite series $\sum_{i=1}^{n} a_i$ there is associated a number, its sum. Even this is not true of every infinite series. The reader who has studied calculus will recall that $\sum_{1}^{\infty} a_i = s$ if and only if $\lim_{n\to\infty} s_n = s$. This means that the number s can be approximated to any degree of accuracy by the nth partial sum s_n for *all* n sufficiently large. In this situation, the series is said to *converge* and its sum is defined to be s. For example

$$\sum_{i=0}^{\infty} (1/10)^i = 10/9 \text{ since } s_n = \sum_{i=0}^{n-1} (1/10)^i = \frac{1-(1/10)^n}{1-1/10}$$
$$= 10/9 - 1/9 \times 10^{-(n-1)}$$

is within $1/9 \times 10^{-(n-1)}$ of 10/9. The number $1/9 \times 10^{-(n-1)}$ can be made arbitrarily small-for instance, if $n > 7$ then $1/9 \times 10^{-(n-1)} < 0.000001$.

On the other hand, the infinite series $1 + 1 + \cdots + 1 + \cdots$ does not converge since $s_n = n$ and so becomes large as n becomes large. Therefore s_n cannot be close to any real number. The series $\sum_{i=1}^{\infty} (-1)^i$ does not have a sum either. Here $s_1 = 1$, $s_2 = 0$, $s_3 = 1$, $s_4 = 0$ and so on. There is no number s which can be approximated by s_n to any degree of accuracy for *all* n sufficiently large since $s_n = 1$ if n is odd and $s_n = 0$ if n is even. Series which do not converge are said to be *divergent*.

In the sequel it will be necessary to consider series of the form

$$\sum_{i=0}^{\infty} a_i x^i = a_0 + a_1 x + a_2 x^2 + \cdots . \tag{8}$$

where the a_i are real numbers and x is a real variable. Such series are called *power series*. The series (8) converges if $x = 0$ to the number a_0. It can be proved that either $x = 0$ is the only point of convergence or else there exists a number $r > 0$ (r may be ∞) such that (8) converges if $-r < x < r$. This number r is called the *radius of convergence* and the set of all real numbers x satisfying $-r < x < r$ is called the *interval of convergence*. This interval is the whole x-axis if $r = \infty$. If the interval of convergence is of finite length, the series may either converge or diverge at $x = r$ and $x = -r$. The series diverges if $x > r$ or $x < -r$. For example, the series $1 + x + x^2 + \cdots$ converges if $-1 < x < 1$. Here $s_n = 1/(1-x) - x^n/(1-x)$. Therefore s_n approximates $1/(1-x)$ as closely as desired for all large n since $R_n = x^n/(1-x)$ can be made arbitrarily small for all large n if $-1 < x < 1$.

The following results on power series will be used. Their proofs can be found in most advanced calculus books.

If $|x|$ is less than the radius of convergence of either of the series $\sum_{n=0}^{\infty} a_n x^n$ and $\sum_{n=0}^{\infty} b_n x^n$, then the *product* of the two series is $\sum_{n=0}^{\infty} c_n x^n$, where $c_n = a_0 b_n + a_1 b_{n-1} + a_2 b_{n-2} + \cdots + a_n b_0$.

Let $f(x) = \sum_{n=0}^{\infty} a_n x^n$ and $g(x) = \sum_{n=0}^{\infty} b_n x^n$ be power series with radii of convergence r and R respectively. The power series $\sum_{k=0}^{\infty} c_k x^k$ which arises formally by substituting the series for $g(x)$ in place of x in the expansion of f and then rearranging terms in increasing powers of x converges under the following condition. If $|x| < R$ and $\sum_{n=0}^{\infty} |b_n x^n| < r$ then for this $x, f(g(x)) = \sum_{k=0}^{\infty} c_k x^k$. The coefficients c_k are obtained as follows: Define the number $b_k(n)$ by the equation

$$g(x)^n = \left(\sum_{k=0}^{\infty} b_k x^k\right)^n = \sum_{k=0}^{\infty} b_k(n)\, x^k.$$

Then $c_k = \sum_{n=0}^{\infty} a_n b_k(n)$ for $k = 0, 1, 2, \ldots$.

Exercises

1 What expressions are represented by the following:

(a) $\sum_{1}^{n} r(r-1)$ (b) $\sum_{n}^{2n} r!$

(c) $\sum_{0}^{n-1} (r+1)\, x^r$ (d) $\sum_{1}^{2n} (-1)^{r+1}\, r^3$

(e) $\sum_{1}^{n} [(r+1)^4 - r^4]$ (f) $\sum_{1}^{n} \frac{1}{r^2}$

(g) $\sum_{k=4}^{8} (k^2 - k + 2)$ (h) $\sum_{0<n<m} (n)^{1/3}$

(i) $\sum_{s \leqslant i \leqslant 81} (\log(1+i) - \log i)$

2 Evaluate the following:

(a) $\prod_{m=2}^{n} \frac{m^2}{m^2 - 1}$ (b) $\prod_{i=-10}^{20} \left(\frac{1}{2}\right)^i$

(c) $\prod_{10<i<25} (i)^{45}$ (d) $\log \prod_{k=1}^{n} k$

(e) $\prod_{0 \leqslant k < 7} \frac{k}{k^2 + 1}$.

3 Use Σ notation to express the sums below:
a) $1 \cdot 3 + 2 \cdot 4 + 3 \cdot 5 + 4 \cdot 6 + \cdots$ n terms
b) $1 - \frac{1}{2} + \frac{1}{3} - \frac{1}{4} + \cdots$ $2r$ terms

c) $\frac{1}{k^3} + \frac{1}{(k+1)^3} + \frac{1}{(k+2)^3} + \cdots + \frac{1}{(2k)^3}$.

4 Use π notation to express the products below:
a) $1 \cdot 3 \cdot 5 \cdot 7 \cdots$ n terms
b) $n!$
c) $1 \cdot 4 \cdot 9 \cdot 13 \cdot 17 \cdots$ k terms

5 Show that $\Sigma\, a_r \Sigma\, b_s = \Sigma\Sigma\, a_r b_s \quad r, s = 1, \ldots, n.$

6 Let $f(x) = \sum_{n=0}^{\infty} a_n x^n$ and $g(x) = \sum_{n=0}^{\infty} b_n x^n$ be power series with radii of convergence r and R respectively. Show that $\alpha f(x) + \beta g(x) = \sum_{n=0}^{\infty} c_n x^n$, where $c_n = \alpha a_n + \beta b_n$, and find the radius of convergence of this power series.

7 a) Show that $\Sigma\, a_r^2 \Sigma\, b_s^2 - (\Sigma\, a_r b_r)^2 = \frac{1}{2}\Sigma\Sigma\, (a_r b_s - a_s b_r)^2 \; r, s = 1, \ldots, n$
b) Deduce that $(\Sigma\, a_r b_r)^2 \leqslant (\Sigma\, a_r^2)\,(\Sigma\, b_r^2)$, (Cauchy's inequality).

2 Binomial theorem

Consider the following expansion:

$$(x_1 + y_1)(x_2 + y_2) = x_1x_2 + x_1y_2 + x_2y_1 + y_1y_2. \tag{1}$$

It is the sum of terms obtained by multiplying together 2 letters, one from each of the two factors, as indicated by the subscripts. Similarly,

$$(x_1 + y_1)(x_2 + y_2)(x_3 + y_3) = x_1x_2x_3 + x_1x_2y_3 + x_2x_3y_1 + x_1x_3y_2$$
$$+ y_1y_2x_3 + y_1y_3x_2 + y_2y_3x_1 + y_1y_2y_3 \tag{2}$$

is the sum of terms of the third degree obtained by multiplying together 3 letters, one from each of the 3 factors.

If in (1) we set $x_1 = x_2 = x$ and $y_1 = y_2 = y$ we obtain

$$(x + y)^2 = x^2 + 2xy + y^2. \tag{1'}$$

Even though the subscripts have been removed each term still arises by multiplying together 2 letters one from each factor. The coefficients of x^2 and y^2 are both 1 since these two terms can only be formed in 1 way. The coefficient of the xy term is 2 since this term can be formed in two ways—namely, by choosing x from the first factor and multiplying it by y from the second or choosing a y from the first factor and multiplying it by x from the second factor. (See (1) for details.)

Setting $x_1 = x_2 = x_3 = x$ and $y_1 = y_2 = y_3 = y$ in (2) we obtain

$$(x + y)^3 = x^3 + 3x^2y + 3xy^2 + y^3. \tag{2'}$$

Again there is only one way to form x^3 and y^3 and so their coefficients are 1. The term x^2y can be formed in 3 ways corresponding to the fact that we must choose a y from one of the 3 factors; the x's come from the remaining factors. Hence the coefficients of x^2y is 3 (cf. equation 2). The coefficient of y^2x is also 3 since we can choose 2 y's from 3 factors in 3 ways. Once the $2y$'s are chosen there is no choice for x; it comes from the remaining factor.

The reader should be able to immediately verify the following expansion:

$$(x + y)^4 = x^4 + \binom{4}{1} x^3y + \binom{4}{2} x^2y + \binom{4}{3} xy^3 + \binom{4}{4} y^4.$$

If some difficulty is experienced in verifying that the coefficients are correct an expansion similar to (2) should be written out.

The following generalization is called the Binomial Theorem* for a positive integral index n.

THEOREM 1

$$(x + y)^n = \sum_{r=0}^{n} \binom{n}{r} x^{n-r}y^r. \tag{3}$$

Proof The expansion of $(x + y)^n$ is the sum of all possible terms of the nth degree in x and y obtained by multiplying the n factors $x + y$ together. Since two numbers in the same factor cannot multiply each other, the term $x^{n-r}y^r$ arises by choosing the y's from r different factors and the x's from the remaining $n - r$ factors. Note that once the factors containing the y's are

---

* The combinatorial symbols $\binom{n}{r}$ are frequently called *binomial coefficients*.

chosen the one's containing the x's are completely determined. Since r factors can be chosen from n in $\binom{n}{r}$ ways the term $x^{n-r}y^r$ will appear $\binom{n}{r}$ times in the sum. Gathering together such similar terms the expansion of $(x + y)^n$ reduces to a sum of terms of the form $\binom{n}{r} x^{n-r}y^r$ for $r = 0, 1, 2, \ldots, n$.

The reader should note the following properties of the binomial expansion.

The expansion of $(x + y)^n$ contains $n + 1$ terms and the $(r + 1)$st term is $\binom{n}{r} x^{n-r}y^r$. This term is called the general term.

The coefficients of terms equidistant from the two ends are equal, for the $(r + 1)$st term from the end is $\binom{n}{n-r} x^r y^{n-r}$, and $\binom{n}{r} = \binom{n}{n-r}$.

If n is even, there is one middle term given by putting $r = \frac{1}{2}n$ in the general term. If n is odd, there are two middle terms given by $r = \frac{1}{2}(n - 1)$ and $r = \frac{1}{2}(n + 1)$.

Example 1

a) $(1 + x)^n = 1 + \binom{n}{1} x + \binom{n}{2} x^2 + \cdots + \binom{n}{r} x^r + \cdots + x^n.$

b) $(1 + x)^5 = 1 + 5x + 10x^2 + 10x^3 + 5x^4 + x^5.$

c)
$$(x - y)^n = x^n - \binom{n}{1} x^{n-1}y + \binom{n}{2} x^{n-2}y^2 + \cdots + (-1)^r \binom{n}{r} x^{n-r}y^r + \cdots + (-1)^n y^n.$$

To prove the last result replace y by $-y$ in (3).

THEOREM 2 *Consider the expansion of* $(1 + x)^n$, *where* x *is positive and* n *is a positive integer.*

(a) *If* $\dfrac{(n + 1)x}{1 + x} = i$ *is an integer than* $\binom{n}{i} x^i = \binom{n}{i-1} x^{i-1}$. *These two terms are greater than any of the other terms.*

(b) *If* $\dfrac{(n + 1)x}{1 + x}$ *is equal to* 1 *plus a positive fraction less than* 1, *then* $\binom{n}{i} x^i$ *is the greatest term.*

Moreover the terms of the expansion increase until the greatest term(s) is (are) reached and then decrease.

Proof Let us denote the rth term in the expansion of $(1 + x)^n$, $\binom{n}{r-1} x^{r-1}$, by a_r. Then

$$\begin{aligned} a_{r+1} &= \frac{n(n-1)\cdots(n-r+1)}{r!} x^r \\ &= \left[\frac{n(n-1)\cdots(n-r+2)}{(r-1)!} x^{r-1}\right]\left[\frac{n-r+1}{r} x\right] \\ &= a_r\left[\frac{n-r+1}{r} x\right]. \end{aligned}$$

Therefore $a_{r+1} > a_r$ if $\frac{n-r+1}{r} x > 1$ or $(n - r + 1)x > r$ or $(n + 1)x > r + rx$ or $\frac{(n+1)x}{1+x} > r$. Therefore

$$a_{r+1} > a_r \quad \text{whenever} \quad r < \frac{(n+1)x}{1+x}, \tag{4}$$

$$a_{r+1} < a_r \quad \text{whenever} \quad r > \frac{(n+1)x}{1+x}, \tag{5}$$

$$a_{r+1} = a_r \quad \text{whenever} \quad r = \frac{(n+1)x}{1+x}, \tag{6}$$

the last statements being proved in a similar manner.

(a) If $\frac{(n+1)x}{1+x} = i$ we see that $a_{i+1} = a_i$ from (6). Letting $r = i - 1$ in (4) we have $a_i > a_{i-1}$; similarly for $r = i - 2$ we have $a_{i-1} > a_{i-2}$. Hence $a_i > a_{i-1} > a_{i-2}$ and so on. Setting $r = i + 1$ in (5) we have $a_{i+1} > a_{i+2}$; similarly for $r = i + 2$, $a_{i+2} > a_{i+3}$. Hence $a_{i+1} > a_{i+2} > a_{i+3}$ and so on. This proves assertion (a) as well as the last statement of the theorem.

(b) Here r, being an integer, can never equal $\frac{(n+1)x}{1+x}$. If $r = i$ then $i < \frac{(n+1)x}{1+x}$ and so $a_{i+1} > a_i$ from (4). If $r = i - 1$ then again $a_i > a_{i-1}$.

Repeating this procedure we obtain $a_{i+1} > a_i > a_{i-1} \cdots > a_1$. If $r = i + 1$ then (5) is valid and we obtain $a_{i+2} < a_{i+1}$. Now let $r = i + 2$; then $a_{i+3} < a_{i+2}$. Repeating this procedure we get $a_{i+1} > a_{i+2} > a_{i+3} \ldots > a_{n+1}$.

COROLLARY *The value of r for which* $\binom{n}{r}$ *is greatest is* $r = \frac{1}{2}n$ *if n is even, and is* $r = \frac{1}{2}(n-1)$ *or* $\frac{1}{2}(n+1)$ *if n is odd.*

The proof follows directly from the theorem upon setting $x = 1$. Examples of the above corollary appear in Pascal's triangle in 1.4.

The greatest term in the expansion of $(x + y)^n$, where x, y are positive and n is a positive integer, can be deduced in the same way from the corresponding relation

$$a_{r+1} = a_r \left(\frac{n - r + 1}{r}\right)\left(\frac{y}{x}\right), \tag{7}$$

where a_r is the rth term in the expansion of $(x + y)^n$. If x or y is negative, this method gives the greatest term in absolute value.

Example 2 Find the greatest term in the expansion $(2 + 4x)^{10}$ when $x = \frac{1}{5}$. If a_r denotes the rth term, (7) implies that $a_{r+1} > a_r$ if

$$\left(\frac{10 - r + 1}{r}\right)\left(\frac{4x}{2}\right) > 1$$

or $(10 - r + 1)\dfrac{2}{5} > r$ since $x = \dfrac{1}{5}$. Therefore

$$a_{r+1} > a_r \quad \text{if} \quad r < 3\tfrac{1}{7}.$$

Accordingly $a_4 > a_3 > a_2 \ldots$ and similarly $a_4 > a_5 > a_6 \ldots$ and so the 4th term is the greatest.

Exercises

1 Write down the expansion of $(x + a)(x + b)(x + c)(x + d)(x + e)$.

2 What is the coefficient of x^4 in the expansion of $\prod_{i=1}^{7}(x + a_i)$?

How many terms are there in the coefficient of x^3?

3 Write down the expansion of:

(a) $(x - 1)^5$ (b) $(x + 2a)^3$ (c) $(x^2 - 3y^2)^4$ (d) $\left(x + \dfrac{1}{x}\right)^6$ (e) $(-x + y^2)^3$.

4 In the expansion of $(6x - 5y)^{19}$ find:
a) the number of terms.
b) the 3rd term, the middle terms, and the 17th term.

5 Find the term independent of x in the expansion of $(5x - 6x^{-4})^9$.

6 Expand in ascending powers of x:
a) $(2 + 2x - x^2)^n$ (as far as x^2).
b) $(1 + x)^n - (1 - x)^n$.
c) $(1 + x)^2 (1 - 2x^3)^8$ (first 3 terms).

7 Expand $\left(x + \frac{1}{x}\right)^3 \left(x - \frac{1}{x}\right)^5$ in descending powers of x.

8 What are the greatest coefficients in the expansions of:
(a) $(1 + x)^{100}$ (b) $(1 + x)^{101}$ (c) $\left(\frac{1}{x} + 3x\right)^n$.

9 Find the greatest terms in the following expansions:
(a) $(1 + 2x)^{11}$ if $x = \frac{1}{3}$. (b) $(8 + x)^{28}$ if $x = \frac{1}{4}$. (c) $(1 + 3x)^7$ if $x = \frac{1}{2}$.

3 Sums involving binomial coefficients

In this section we shall denote ${}_nC_r$ by C_r, unless n varies. Finite sums of the binomial coefficients C_r are usually evaluated by combinations and variations of the following 6 methods.*

Method I Interpret the sum combinatorially.
An example appears in equation (1) of 1.6.

Method II Substitute a given value for x in a simple expression which expanded by the binomial theorem coincides with the given finite sum.

Example 1

a) $C_0 + C_1 + \cdots + C_n = 2^n$.

b) $C_0 + C_2 + C_4 + \cdots = C_1 + C_3 + C_5 + C_7 + \cdots = 2^{n-1}$.

(a) In the expansion of $(1 + x)^n$ set $x = 1$ obtaining

$$(1 + 1)^n = C_0 + C_1 + \cdots + C_n.$$

(b) In the expansion of $(1 + x)^n$ set $x = -1$ obtaining

$$(1 - 1)^n = C_0 - C_1 + C_2 - \cdots + (-1)^r C_r + \cdots + (-1)^n C_n.$$

* The reader familiar with mathematical induction will see that it is also applicable.

Therefore

$$C_0 + C_2 + C_4 + \cdots = C_1 + C_3 + C_5 + \cdots;$$

let s be the common sum of these two series.

By (a), $2S = 2^n$ or $S = 2^{n-1}$.

Method III Express the given series as a sum of 2 or more binomial expansions.

Example 2 Sum the series $S = C_0 + 2C_1 + 3C_2 + \cdots + (n+1)\,C_n$.

The given series can be summed by finding a simple expression for $C_0 + 2C_1x + 3C_2x^2 + \cdots + (n+1)\,C_nx^n$ and setting $x = 1$. This last series can be written in the form

$$(C_0 + C_1x + C_2x^2 + \cdots + C_nx^n) + (C_1x + 2C_2x^2 + \cdots + nC_nx^n).$$

The first parenthesis is equal to $(1+x)^n$. The second parenthesis, written in more detail, is

$$nx + 2\,\frac{n(n-1)}{2!}\,x^2 + 3\,\frac{n(n-1)(n-2)}{3!}\,x^3 + \cdots + nx^n$$

$$= nx\left(1 + (n-1)\,x + \frac{(n-1)(n-2)}{2!}\,x^2 + \cdots + x^{n-1}\right)$$

$$= nx\,(1+x)^{n-1}.$$

Therefore S is the value of $(1+x)^n + nx\,(1+x)^{n-1}$ when $x = 1$; that is, $S = 2^n + n2^{n-1}$. It is easier to work example 2 by using the next method.

Method IV Obtain the series by differentiation of a finite series whose sum is known.*

Example 3 Sum the series

$$S = C_1 - 2^2C_2 + 3^2C_3 - \cdots + (-1)^{n-1}\,n^2C_n \quad (n > 2).$$

Differentiating both sides of the following identity

$$C_0 + C_1x + C_2x^2 + C_3x^3 + \cdots + C_nx^n = (1+x)^n,$$

yields

$$C_1 + 2C_2x + 3C_3x^2 + \cdots + nC_nx^{n-1} = n\,(1+x)^{n-1}.$$

---

* Methods IV and V may be omitted by students unfamiliar with calculus.

Multiply both sides of the above identity by x and then differentiate with respect to x. We obtain

$$C_1 + 2^2 C_2 x + 3^2 C_3 x^2 + \cdots + n^2 C_n x^{n-1}$$
$$= n(1 + x)^{n-1} + n(n - 1)x(1 + x)^{n-2}.$$

Therefore setting $x = -1$, we see that the left side of the above identity reduces to S while the right side equals 0 and so $S = 0$.

Method V Obtain the series by integration of a finite series whose sum is known.

Example 4 Prove that $C_1 - \frac{1}{2} C_2 + \frac{1}{3} C_3 - \cdots + (-1)^n \frac{1}{n} C_n = 1 + \frac{1}{2} + \frac{1}{3} + \cdots + \frac{1}{n}$.

Since $1 + C_1 x + C_2 x^2 + \cdots + C_n x^n = (1 + x)^n$,

$$C_1 + C_2 x + C_3 x^2 + \cdots + C_n x^{n-1} = \frac{(1 + x)^n - 1}{x}.$$

Integrating from $x = -1$ to $x = 0$ yields

$$C_1 - \frac{1}{2} C_2 + \frac{1}{3} C_3 + \cdots + (-1)^{n-1} \frac{1}{n} C_n = \int_{-1}^{0} \frac{(1 + x)^n - 1}{x} \, dx.$$

Setting $1 + x = y$ in the last integral yields

$$\int_0^1 \frac{y^n - 1}{y - 1} \, dy = \int_0^1 (1 + y + y^2 + \cdots + y^{n-1}) \, dy = 1 + \frac{1}{2} + \frac{1}{3} + \cdots + \frac{1}{n}.$$

Method VI Build up a function in which the given series is the coefficient of a particular power of the variable and evaluate this coefficient in an entirely different manner.

Example 5 Find the value of $C_0^2 + C_1^2 + C_2^2 + \cdots + C_n^2 = S$.
We use the identity

$$(C_0 + C_1 x + C_2 x^2 + \cdots + C_n x^n)(C_0 + C_1 x + \cdots + C_n x^n)$$
$$= (1 + x)^n (1 + x)^n.$$

The coefficient of x^n on the left side is

$$C_0C_n + C_1C_{n-1} + C_2C_{n-2} + \cdots + C_nC_0 = C_0^2 + C_1^2 + C_2^2 + \cdots + C_n^2,$$

since $C_r = C_{n-r}$ $(r = 0, 1, 2, \ldots, n)$. Therefore S equals the coefficient of x^n in $(1 + x)^{2n}$ which is $\binom{2n}{n} = \frac{(2n)!}{n!\,n!}$.

Exercises

1 Find the value of $C_0C_1 + C_1C_2 + C_2C_3 + \cdots + C_{n-1}C_n$.

2 Prove that

$$(1 + C_1 + C_2 + \cdots + C_n)^2 = 1 + \binom{2n}{1} + \binom{2n}{2} + \cdots + \binom{2n}{2n}.$$

3 Show that for $n \geqslant 2$

a) $C_1 - 2C_2 + 3C_3 - \cdots (-1)^{n-1}\, nC_n = 0.$

b) $2 \cdot 1C_2 + 3 \cdot 2C_3 + 4 \cdot 3C_4 + \cdots + n\,(n-1)\,C_n = n\,(n-1)\,2^{n-2}.$

4 By comparison of the coefficients of t^n on both sides of

$$(1 + t)^a\,(1 + t)^b = (1 + t)^{a+b}$$

show that

$$\binom{a}{0}\binom{b}{n} + \binom{a}{1}\binom{b}{n-1} + \cdots + \binom{a}{n}\binom{b}{0} = \binom{a+b}{n}.$$

5 Show that $\sum_{i=0}^{n} (-1)^i \binom{a}{i} = (-1)^n \binom{a-1}{n}$ by using equation 1.4.3.

6 Prove that $C_0 + \frac{1}{2}\,C_1 + \frac{1}{3}\,C_2 + \cdots + \frac{1}{n+1}\,C_n = \frac{1}{n+1}\,(2^{n+1} - 1).$

7 What is the value of $C_0 - 2C_1 + 3C_2 - \cdots + (-1)^n\,(n+1)\,C_n$?

8 Prove that $C_1 + 3C_3 + 5C_5 + \cdots = 2C_2 + 4C_4 + 6C_6 \cdots = n \cdot 2^{n-2}.$

9 Obtain relations by equating the coefficients of x^k in the expansion of each side of the following identities:

a) $(1 + x)^{n+1} = (1 + x)\,(C_0 + C_1x + \cdots + C_nx^n).$

b) $(1 + x)^{n+2} = (1 + 2x + x^2)\,(C_0 + C_1x + \cdots + C_nx^n).$

c) $(1 + x)^{n+3} = (1 + 3x + 3x^2 + x^3)\,(1 + x)^n.$

10 Prove Vandermonde's identity:

$$\binom{m}{r} + \binom{m}{r-1}\binom{n}{1} + \binom{m}{r-2}\binom{n}{2} + \cdots + \binom{n}{r} = \binom{m+n}{r}$$

by the use of the identity $(1 + x)^m\,(1 + x)^n = (1 + x)^{m+n}$.

11 Prove that $C_0 + 3C_1 + 5C_2 + \cdots + (2n+1)\,C_n = (n+1)\,2^n.$

12 Prove that $C_2 + 2C_3 + 3C_4 + \cdots + (n-1)\,C_n = 1 + (n-2)\,2^{n-1}.$

4 The multinomial theorem

The multinomial theorem for positive integral n is just a generalization of theorem 2.1 when more than two variables appear in each factor. For example,

$$(x_1 + x_2 + x_3)^3 = x_1^3 + x_2^3 + x_3^3 + 3x_1^2x_2 + 3x_1^2x_3 + 3x_2^2x_1 + 3x_3^2x_2 + 3x_3^2x_1 + 6x_1x_2x_3 = \sum \frac{3!}{i_1!\,i_2!\,i_3!}\, x_1^{i_1}x_2^{i_2}x_3^{i_3}$$

where $\sum$ indicates summation over all nonnegative integers i_1, i_2, i_3 such that $i_1 + i_2 + i_3 = 3$. To verify the last equality we note that all possible values of (i_1, i_2, i_3) are: (3, 0, 0), (0, 3, 0), (0, 0, 3), (2, 1, 0), (2, 0, 1), (0, 2, 1), (1, 2, 0), (0, 1, 2), (1, 0, 2) and (1, 1, 1).

The triplets of numbers listed above are called *ordered 3-partitions of 3*. The term *3-partition* of 3 refers to the fact that 3 nonnegative integers are added together to give 3. *Ordered* refers to the fact that the same combination of elements in a different order is counted as a different partition. If order is ignored then there are only three 3-partitions of 3, namely, (0,0,3), (2, 0, 1) and (1, 1, 1). We will only consider ordered partitions in this section and so will omit the word "ordered". Some 2-partitions of 8 are (2, 6), (6, 2), (4, 4), (5, 3), and (3, 5). A 4-partition of 6 is (1, 1, 1, 3). As an exercise the reader should list all fifteen 3-partitions of 4 such as (4, 0, 0) etc. In general an *ordered k-partition of n* is the set of all ordered k-tuples $(i_1, i_2, \ldots, i_k)$ of nonnegative integers such that $i_1 + i_2 + \cdots i_k = n$.

THEOREM 1

$$(x_1 + x_2 + \cdots + x_k)^n = \sum \frac{n!}{i_1!\,i_2!\cdots i_k!}\, x_1^{i_1} \cdots x_k^{i_k}$$

where the summation extends over all nonnegative integers $(i_1, i_2, \ldots, i_k)$ *such that* $i_1 + i_2 + \cdots + i_k = n$ (ordered k-partitions of n).

Proof The expansion of $(x_1 + x_2 + \cdots + x_k)^n$ is the sum of all possible terms of the nth degree in $x_1, x_2, \ldots, x_k$, obtained by multiplying the n factors $(x_1 + x_2 + \cdots + x_k)$ together. Since two numbers in the same factor cannot multiply each other, the term $x_1^{i_1}x_2^{i_2}x_3^{i_3}\cdots x_k^{i_k}$ arises by choosing the x_1's from i_1 different factors, the x_2's from another i_2 different factors, the x_3's from still another different set of factors and so on. In other words, we divide the n factors into k groups containing $i_1, i_2, \ldots, i_k$ factors. From the first group containing i_1 factors we choose the x_1's, from the second group containing i_2 factors we choose the x_2's etc. By theorem 1.5.1 this division can be carried out in $\dfrac{n!}{i_1!\,i_2!\cdots i_k!}$ ways. For each of these ways we obtain the term $x_1^{i_1}x_2^{i_2}\cdots x_k^{i_k}$. Collecting such similar terms, the expansion of $(x_1 + x_2 + \cdots + x_k)^n$ reduces to a sum of terms of the form

$$\frac{n!}{i_1!\,i_2!\cdots i_k!}\,x_1^{i_1}x_2^{i_2}\cdots x_k^{i_k},$$

where the summation extends over all nonnegative integers $i_1, i_2, \ldots, i_k$ such that $i_1 + i_2 + \cdots + i_k = n$.

COROLLARY $k^n = \sum \dfrac{n!}{i_1!\,i_2!\cdots i_k!}$, *where the summation extends over all nonnegative integers* $i_1, i_2, \ldots, i_k$ *such that* $i_1 + i_2 + \cdots + i_k = n$.

The proof follows immediately from the multinomial expansion by setting $x_1 = x_2 = \cdots = x_k = 1$.

Example 1 Expand $(x - 3y - 2z)^3$.

To expand the above expression, set $x_1 = x$, $x_2 = -3y$ and $x_3 = -2z$ and use the expansion of $(x_1 + x_2 + x_3)^3$. Hence $(x - 3y - 2z)^3 = x^3 - 27y^3 - 8z^3 - 9x^2y - 6x^2z - 54y^2z + 27y^2x - 36z^2y + 12z^2x + 36xyz$.

Example 2 Find the coefficient of $xy^2z^3w^2$ in the expansion of $(x - y - z + w)^8$. The term $x(-y)^2(-z)^3w^2 = -xy^2z^3w^3$ has as coefficient $\dfrac{8!}{1!\,2!\,3!\,2!}$. Therefore the coefficient of $xy^2z^3w^2$ is $-\dfrac{8!}{2!\,3!\,2!}$.

Exercises

1 Write down the expansion of:
(a) $(x - y - 2z)^2$. (b) $(x - 3y + 5z)^3$. (c) $(2x - 3y + z)^4$.

2 Find the coefficient of $x_1^2x_2^3x_3^4x_5^5$ in $(-2x_1 + 3x_2 - 4x_3 - x_4 + x_5)^{15}$.

3 Find the coefficient of $x_1^{10}x_2^{11}x_3^4$ in $(x_1 - 7x_2 - 2x_3)^{25}$.

4 Interpret the corollary to the theorem combinatorially (c.f. equation (1) of 1.6).

5 How many terms are there in the expansion of $(x_1 + x_2 + \cdots + x_k)^n$ or in other words, how many different ordered solutions are there of the equation $i_1 + i_2 + \cdots + i_k = n$ where each i is a nonnegative integer?

5 The binomial series

Suppose we expand $(1 - x)^{-1}$ by means of the binomial expansion

$$(1 + x)^n = 1 + nx + \frac{n(n-1)}{2!} x^2 + \cdots + \frac{n(n-1)(n-2)\cdots(n-r+1)}{r!} x^r + \cdots \tag{1}$$

by setting $n = -1$ and formally calculating the coefficients. We obtain the series

$$\frac{1}{1-x} = 1 + x + x^2 + x^3 + \cdots \tag{2}$$

which is valid if $|x| < 1$ (see section 1).

We recall that we have only proven the binomial theorem when n is a nonnegative integer. However, encouraged by the above example, we could try letting n be any real number. The series we obtain in these cases by using (2), does not terminate.

Example 1

a) $(1 + x)^{-1} = 1 - x + x^2 - x^3 + x^4 - \cdots.$

b) $(1 + x)^{-1/2} = 1 - \frac{1}{2} x + \frac{1 \cdot 3}{2 \cdot 4} x^2 - \frac{1 \cdot 3 \cdot 5}{2 \cdot 4 \cdot 6} x^3 + \cdots.$

c) $(1 + x)^{1/2} = 1 + \frac{1}{2} x + \frac{(-1)(1)}{2 \cdot 4} x^2 - \frac{(-1)(1)(3)}{2 \cdot 4 \cdot 6} x^3 + \cdots.$

Series of the above form are called *binomial series*. We have seen that they do not, in general, possess "sums to infinity" for all values of x. This problem was not encountered for a nonnegative integer n since (2) terminates automatically after $n + 1$ terms as each of the later coefficients contains the factor $n - n = 0$.

The proof of the following theorem can be found in most calculus books.

THEOREM 1 *If n is any real number, and if* $|x| < 1$, *the sum to infinity of*

$$1 + nx + \frac{n(n-1)}{2!}x^2 + \cdots + \frac{n(n-1)(n-2)\cdots(n-r+1)}{r!}x^r + \cdots$$

is the positive *value of* $(1+x)^n$.

Observe that the function $(a+x)^n$ can have two values; we shall use the symbol $(a+x)^n$ to denote only the positive value. For example, if $x = \frac{1}{2}$ then $(1+x)^{1/2} = \sqrt{\frac{3}{2}}$ and not $-\sqrt{\frac{3}{2}}$.

The binomial coefficients $\binom{n}{r}$ have only been used when n and r are positive integers. Theorem 1 shows that it is convenient to extend this definition. The *general binomial coefficients* are defined by

$$\binom{x}{r} = \frac{x(x-1)\cdots(x-r+1)}{r!}$$

for all values of x and all positive integers r. For $r = 0$, $\binom{x}{0} = 1$ and for negative integers r, $\binom{x}{r} = 0$. The symbol $\binom{x}{r}$ will not be used if r is not an integer. It may easily be verified from the definition that $\binom{n}{r} = 0$ if $r > n$ and that

$$\binom{x}{r-1} + \binom{x}{r} = \binom{x+1}{r}$$

for any number x and any integer r.

The expansion of $(a+x)^n$ can be obtained from theorem 1. If $\left|\frac{x}{a}\right| < 1$ and if a^n exists,

$$(a+x)^n = a^n\left(1+\frac{x}{a}\right)^n = a^n\left[1 + n\left(\frac{x}{a}\right) + \frac{n(n-1)}{2!}\left(\frac{x}{a}\right)^2 + \cdots\right].$$

However if $\left|\frac{x}{a}\right| > 1$, we cannot expand the function in ascending powers of x but if x^n exists,

$$(a+x)^n = x^n\left(1+\frac{a}{x}\right)^n = x^n\left(1 + n\frac{x}{a} + \frac{n(n-1)}{2!}\frac{a^2}{x^2} + \cdots\right).$$

This is an expansion in descending powers of x.

Example 2 Expand $(1+4x)^{5/3}$ given $|x| < \frac{1}{4}$.

The function can be expanded in ascending powers of x if $|4x| < 1$ or $|x| < \frac{1}{4}$. The expansion is

$$1 + \tfrac{5}{3}(4x) + \frac{(\frac{5}{3})(\frac{2}{3})}{1\cdot 2}(4x)^2 + \frac{(\frac{5}{3})(\frac{2}{3})(-\frac{1}{3})}{1\cdot 2\cdot 3}(4x)^3 + \cdots$$

$$= 1 + \frac{5}{1}\left(\frac{4}{3}x\right) + \frac{5\cdot 2}{1\cdot 2}\left(\frac{4x}{3}\right)^2 + \frac{5\cdot 2\cdot(-1)}{1\cdot 2\cdot 3}\left(\frac{4x}{3}\right)^2 + \cdots$$

The terms which follow are alternately positive and negative and the general term is

$$(-1)^r \frac{5\cdot 2\cdot 1\cdot 4\cdot 7\cdots(3r-8)}{r!}\left(\frac{4x}{3}\right)^r, \quad r > 2.$$

Example 3 Expand $(1-x)^{-4}$ given $|x| < 1$. Find the greatest term(s) in the expansion if $x = \frac{1}{2}$.

$$(1-x)^{-4} = 1 + (-4)(-x) + \frac{(-4)(-5)}{1\cdot 2}(-x)^2 + \frac{(-4)(-5)(-6)}{1\cdot 2\cdot 3}(-x)^3 + \cdots$$

$$= 1 + 4x + \frac{4\cdot 5}{1\cdot 2}x^2 + \frac{4\cdot 5\cdot 6}{1\cdot 2\cdot 3}x^3 + \cdots$$

The $(r+1)$st term, a_{r+1}, is given by

$$a_{r+1} = \frac{(-4)(-5)(-6)\cdots(-r-3)}{r!}(-x)^r = \frac{(r+1)(r+2)(r+3)}{1\cdot 2\cdot 3}x^r.$$

Therefore

$$\frac{a_{r+1}}{a_r} = \frac{r+3}{r}x.$$

If $x = \frac{1}{2}$, $a_{r+1} > a_r$ as long as $r+3 > 2r$ or $r < 3$. Hence $a_2 > a_1$ and $a_3 > a_2$ or $a_1 < a_2 < a_3$. Setting $r = 3$ yields $a_3 = a_4$. If $r > 3$, $a_{r+1} < a_r$ and so $a_5 < a_4$, $a_6 < a_5$ and so forth. Therefore $a_4 > a_5 > a_6 \cdots$. Hence there are two terms $a_3 = a_4$ which are greater than all other terms.

Example 4 Find the coefficient of x^{15} in $(x^2 + x^3 + x^4 + x^5) \times (x + x^2 + \cdots + x^7)(1 + x + \cdots + x^{15})$.

The above expression can be reduced to

$$x^3(1 + x + x^2 + x^3)(1 + x + \cdots + x^6)(1 + x + \cdots + x^{15})$$
$$= x^3\left(\frac{1 - x^4}{1 - x}\right)\left(\frac{1 - x^7}{1 - x}\right)\left(\frac{1 - x^{16}}{1 - x}\right).$$

Hence we need only find the coefficient of x^{12} in

$$(1 - x^4)(1 - x^7)(1 - x^{16})(1 - x)^{-3}$$

which is the same as the coefficient of x^{12} in

$$(1 - x^4)(1 - x^7)(1 - x)^{-3}$$

since the term x^{16} can be neglected. The above expression reduces to

$$(1 - x^4 - x^7 + x^{11})(1 - x)^{-3}.$$

The coefficient of x^{12} equals the coefficient of x^{12} in $(1 - x)^{-3}$ minus the coefficient of x^8 in $(1 - x)^{-3}$ minus the coefficient of x^5 in $(1 - x)^{-3}$ plus the coefficient of x in $(1 - x)^{-3}$. The remainder of the calculation is left to the reader.

Exercises

1 Write down the first 3 terms and the coefficient of x^r in the expansion in ascending powers of x of the following functions. In each case, state when the expansion is valid.

(a) $(1 - 5x)^{-1}$ (b) $(1 - 2x)^{-2}$ (c) $(1 + x)^{-3}$ (d) $(1 + 2x)^{-4}$ (e) $(1 + 4x)^{3/2}$

(f) $x(1 + x^2)^{-1}$

2 Find the coefficient of x^3 in the expansion of $(1 + 3x)^{-1/3}(1 + 2x)^{1/2}$.

3 Find the coefficient of x^{14} in the expansion of $(1 + x + x^2 + \cdots + x^8)^{10}$.

4 Find the indicated terms in the following expansions.
a) 4th term in $(1 + 3x)^{5/2}$ if $|x| < 1/3$.
b) 6th term in $(1 - 3x^2)^{-10}$ if $|x| < 1/2$.
c) 10th term in $(1 - 4x)^{-2}$ if $|x| < 1/4$.
d) 8th term in $(1 - 5x)^{-3}$ if $|x| < 1/5$.
e) 5th term in $(1 + 4x^2)^{-4/3}$ if $|x| < 1/2$.

5 Let a_r denote the r-th term in the binomial series (1).

a) show that $\dfrac{a_{r+1}}{a_r} = \dfrac{n - r + 1}{r}x = -\left(1 - \dfrac{n + 1}{r}\right)x.$

b) If $r > n + 1$ prove that $\dfrac{a_{r+1}}{a_r}$ is negative if $x > 0$ and positive if $x < 0$.

c) Deduce that after a certain stage, the terms are alternatively positive and negative if $x > 0$, and of the same sign if $x < 0$.

6 How many terms are positive in the expansion of $(1 - x^2)^{5/4}$ where $|x| < 1$?

7 How many terms are negative in the expansion of $(1 - x^2)^{15/2}$ where $|x| < 1$?

8 Generalize theorem 2.2 so that it applies to the binomial series.

9 Which is the numerically greatest term (or terms) in the expansion of the following functions?

a) $(1 + x)^{13/2}$ when $x = 2/3$
b) $(1 + x)^{-1/2}$ when $x = 1/4$
c) $(1 - x)^{-10}$ when $x = 3/4$
d) $(1 - x)^{2/3}$ when $x = 1/10$.

10 a) Verify that

$$(1 - x)^{-n} = 1 + \frac{n}{1} x + \frac{n(n+1)}{1 \cdot 2} x^2 + \frac{n(n+1)(n+2)}{1 \cdot 2 \cdot 3} x^3 + \cdots.$$

$$(1 + x)^{-n} = 1 - \frac{n}{1} x + \frac{n(n+1)}{1 \cdot 2} x^2 - \frac{n(n+1)(n+2)}{1 \cdot 2 \cdot 3} x^3 + \cdots.$$

b) Replace b by r/s in (a) and show that

$$(1 - x)^{-r/s} = 1 + \frac{r}{1}\left(\frac{x}{s}\right) + \frac{r(r+s)}{1 \cdot 2}\left(\frac{x}{s}\right)^2 + \frac{r(r+s)(r+2s)}{1 \cdot 2 \cdot 3}\left(\frac{x}{s}\right)^3 + \cdots$$

c) Rewrite the series

$$1 + \frac{10}{9} + \frac{10 \cdot 16}{9 \cdot 18} + \frac{10 \cdot 16 \cdot 22}{9 \cdot 18 \cdot 27} + \cdots$$

in the form

$$1 + \frac{5/3}{1}\left(\frac{6}{9}\right) + \frac{(5/3)(8/3)}{1 \cdot 2}\left(\frac{6}{9}\right)^2 + \frac{(5/3)(8/3)(11/3)}{1 \cdot 2 \cdot 3}\left(\frac{6}{9}\right)^3 + \cdots$$

and apply part (b) to show that the sum to infinity of the above series is $(1 - 2/3)^{-5/3}$.

11 Find the sum to infinity of the following series:

a) $1 + 2(1/2) + 3(1/2)^2 + 4(1/2)^3 + \cdots$

b) $1 + 6(4/5) + \dfrac{6 \cdot 7}{1 \cdot 2}\left(\dfrac{4}{5}\right)^2 + \dfrac{6 \cdot 7 \cdot 8}{1 \cdot 2 \cdot 3}\left(\dfrac{4}{5}\right)^3 + \cdots$

c) $1 - \dfrac{1}{4} + \dfrac{1 \cdot 3}{4 \cdot 8} - \dfrac{1 \cdot 3 \cdot 5}{4 \cdot 8 \cdot 12} + \cdots$

d) $\dfrac{1}{24} - \dfrac{1 \cdot 3}{24 \cdot 32} + \dfrac{1 \cdot 3 \cdot 5}{24 \cdot 32 \cdot 40} - \cdots$

12 Expand $\dfrac{x^3 + x^2 + 6}{x^2 - 1}$ in ascending powers of x. (Hint: Write the function as a partial fraction.)

13 Expand $\dfrac{3x+3}{(x-1)(x^2+x+1)}$ in ascending powers of x. Repeat the problem for $\dfrac{5x+5}{(x-1)(x+4)}$

14 a) Evaluate $\sqrt{2}$ to 6 places of decimals by using the identity $\dfrac{7\sqrt{2}}{10} = (1 - 2.0)^{1/2}$ and expanding the right side.

b) Show that only 4 terms are required.

15 a) Expand $(1+x)^r$ and $(1+x)^s$ by using Theorem 1. Multiply these two series together and show that the coefficient of x^m is $\sum_{i=0}^{m} \binom{r}{i}\binom{s}{m-i}$.

b) Verify that the coefficient of x^m in the expansion of $(1+x)^{r+s}$ is $\binom{r+s}{m}$ and then by using (a) and the identity $(1+x)^r(1+x)^s = (1+x)^{r+s}$ show that

$$\left(\sum_{i=0}^{m}\right)\binom{r}{i}\binom{s}{m-i} = \binom{r+s}{m}.$$

c) Give an alternative proof of the last identity by considering the number of ways of choosing m balls from an urn containing r red balls and s black balls where each ball is a different size.

CHAPTER III

Generating functions

1 Combinatorial generating functions

EACH THEOREM in chapter I required a different method of proof. These different techniques of enumeration can be unified by the use of generating functions.

To introduce this new concept, we describe an algebraic process which automatically produces r-combinations. We begin by considering two objects x_1 and x_2. Consider the product

$$(1 + x_1t)(1 + x_2t) = 1 + (x_1 + x_2)t + (x_1x_2)t^2. \tag{1}$$

The coefficient of t is just the sum of all possible 1-combinations while the coefficient of t^2 is just the 2-combination x_1x_2. Suppose there are 3 objects x_1, x_2, and x_3. Expanding the product

$$(1 + x_1t)(1 + x_2t)(1 + x_3t) \tag{2}$$

in powers of t yields

$$1 + (x_1 + x_2 + x_3)t + (x_1x_2 + x_1x_3 + x_2x_3)t^2 + (x_1x_2x_3)t^3. \tag{3}$$

Again we observe that the coefficient of t is the sum of all possible 1-combinations, the coefficient of t^2 is the sum of all possible 2-combinations and the coefficient of t^3 is the unique 3-combination. To obtain the number of 2-combinations, we need only count the number of terms in the coefficient of t^3 in (3). If we are only interested in the number of terms we can set $x_1 = x_2 = x_3 = 1$; then the coefficient of t^2 is 3 which is the number of terms. Actually, we can find the coefficient of any term in (3) directly from (2). Setting $x_1 = x_2 = x_3 = 1$, (2) reduces to $(1 + t)^3$. The binomial theorem immediately gives us the coefficient of any term.

This process can be generalized to the case of n distinct things $x_1, x_2, \ldots, x_n$. We have

$$\prod_{i=1}^{n} (1 + x_it) = 1 + s_1t + s_2t^2 + \cdots + s_nt^n \tag{4}$$

where

$$s_0(x_1, x_2, \ldots, x_n) = 1.$$

$$s_1(x_1, x_2, \ldots, x_n) = x_1 + x_2 + \cdots + x_n.$$

$$s_2(x_1, x_2, \ldots, x_n) = x_1x_2 + x_1x_3 + \cdots + x_1x_n + \cdots + x_{n-1}x_n.$$

$$s_3(x_1, x_2, \ldots, x_n) = x_1x_2x_3 + x_1x_2x_4 + \cdots + x_{n-2}x_{n-1}x_n.$$

....

$$s_n(x_1, x_2, \ldots, x_n) = x_1x_2x_3 \ldots x_n.$$

The s_i $(i = 1, \ldots, n)$ are called the *elementary symmetric functions* of n variables. Each s_i is the sum of all i-combinations formed from the n objects $x_1, x_2, \ldots, x_n$. Hence

$$s_i(1, 1, 1, \ldots, 1) = \binom{n}{i} \qquad i = 0, 1, 2, \ldots, n.$$

Setting $x_1 = x_2 = \cdots = x_n = 1$, (4) reduces to

$$(1 + t)^n = 1 + \binom{n}{1} t + \binom{n}{2} t^2 + \cdots + \binom{n}{n} t^n.$$

The function $(1 + t)^n$ is the *generating function* for the sequence of numbers $\binom{n}{0}, \binom{n}{1}, \ldots, \binom{n}{n}$. In general, given a sequence of numbers a_0, $a_1, a_2, \ldots, a_n, \ldots$ the function $f(t)$ defined by

$$f(t) = \sum_{i=0}^{\infty} a_i t^i$$

is called the *enumerating* or *generating function* of the sequence*. Examples of generating functions for finite sequences can be found in chapter II section 3. The methods of summation listed there are all based upon adding, multiplying, differentiating, and integrating suitable generating functions. Additional examples are given in the table below.

---

* The variable t can be interpreted as an abstract indeterminate. The existence of the sum is then a matter of a formal definition. Sums and products of such series always exist. This formal view point appears in E.T.Bell, "Euler algebra", *Tran. Amer. Math. Soc.*, **25** 135–154 (1923).

a_k	$f(t)$
$\binom{n}{k}$	$(1 + t)^n$
1	$(1 - t)^{-1}$
k	$t(1 - t)^{-2}$
k^2	$t(1 + t)(1 - t)^{-3}$
$(-1)^k$	$(1 + t)^{-1}$
x^k	$(1 - xt)^{-1}$

Let us now examine the reason for the appearence of the symmetric functions as the coefficients of powers of t. To obtain the coefficients of t we must add all terms which contain t. A t is obtained iff a single t is chosen from any of the n factors and a 1 from the remaining $(n - 1)$ factors. When t is chosen from the ith factor the term $x_i t$ is obtained. Hence $x_1 + x_2 + \cdots + x_n$ will appear as the coefficient of t. Similarly, the coefficient of t^2 is obtained when all terms containing t^2 are added. A t^2 is obtained iff a t is chosen from any two of the n factors and the 1's from the remaining factors. If the ith and jth factors are chosen, the term $x_i x_j t^2$ is obtained. Hence the coefficient of t^2 will be s_2. Since a t^r is obtained iff a t is chosen from any r of the n factors and ones are chosen from the remaining factors, the coefficient of t is s_r, the sum of all possible products of degree r (r-combinations) which can be formed from the n letters $x_1, x_2, \ldots, x_n$.

Suppose any of the n distinct letters $x_1, x_2, \ldots, x_n$ can appear from 0 to r times in an r-combination as in theorem 1.4.2. How do we find the generating function for r-combinations of this type? We recall that each r-combination corresponds uniquely to a homogeneous product of degree r and conversely. The homogeneous product is formed by taking all the letters which appear in the r-combination and multiplying them together. For instance, if $n = 4$ and $r = 3$ the r-combination (x_1, x_1, x_1), (x_1, x_2, x_3), (x_4, x_4, x_1) are represented by x_1^3, $x_1x_2x_3$, and $x_1x_4^2$ respectively. The advantage of this interpretation is that homogeneous products can be constructed algebraically. Consider the product

$$\begin{aligned}(1 + x_1t + x_1^2t^2 + x_1^3t^3 + \cdots)(1 + x_2t + x_2^2t^2 + x_2^3t^3 + \cdots)\\ \cdots(1 + x_nt + x_n^2t^2 + x_n^3t^3 + \cdots). \qquad (5)\end{aligned}$$

Multiplied out and arranged in powers of t, (5) becomes

$$\begin{aligned}1 + (x_1 + x_2 + \cdots + x_n)\,t + (x_1^2 + x_2^2 + \cdots + x_n^2 + x_1x_2 + \cdots + x_1x_n\\ + x_2x_3 + \cdots + x_2x_n + \cdots + x_{n-1}x_n)\,t^2 + \cdots. \qquad (6)\end{aligned}$$

Note that the coefficients of t are homogeneous products of degree 1 while those of t^2 are homogeneous products of degree 2. To obtain t^r we can select t^{i_1} from the first factor in (5), t^{i_2} from the second factor, ..., t^{i_n} from the nth factor, where $i_1 + i_2 + \cdots + i_n = r$ the i's being nonnegative integers. The coefficient of t^r for this choice of powers of t is $x_1^{i_1} x_2^{i_2} \cdots x_n^{i_n}$. Adding together all such terms subject only to the restriction, $i_1 + i_2 + \cdots + i_n = r$, we see that the coefficient of t^r is just the sum of the homogeneous products of degree r formed from the n letters $x_1, x_2, \ldots, x_n$.

To count the number of homogeneous products of degree r we need only set $x_1 = x_2 = \cdots = x_n = 1$ and find the coefficient of t^r in (6). This coefficient can be found by expanding (5) which reduces to

$$(1 + t + t^2 + \cdots + t^m + \cdots)^n = \left(\frac{1}{1-t}\right)^n = (1-t)^{-n}.$$

Applying theorem 2.5.1 yields

$$\begin{aligned}(1-t)^{-n} &= \sum_{r=0}^{\infty} (-1)^r \frac{(-n)(-n-1)(-n-2)\cdots(-n-r+1)}{1\cdot 2\cdot 3\cdots r} t^r \\ &= \sum_{r=0}^{\infty} \frac{n(n+1)(n+2)\cdots(n+r-1)}{r!} t^r \\ &= \sum_{t=0}^{\infty} \binom{n+r-1}{r} t^r. \qquad (7)\end{aligned}$$

Therefore $(1-t)^{-n}$ is the generating function for the combinations formed from n distinct objects with no restriction on the number of times an object may appear (cf. theorem 1.4.2).

The above examples serve to illustrate that the rules of multiplication automatically ensure the formation of r-combinations which are added together and appear as coefficients of t^r. The reason for this is that each term of the product is formed by independently choosing a letter from each factor. This is exactly what we do in forming a single r-combination. Suppose an element x_1 can appear 0 or 1 or 2, ..., or k times. Since the multiplication process automatically picks out each element in a factor and we want x_1 to be chosen 0 or 1 or 2 or $\cdots$ or k times its factor should be

$$(1 + x_1 t + x_1^2 t^2 + \cdots + x_1^k t^k).$$

The ith factor is a generating function for the sequence $x_i, x_i^2, x_i^3, \ldots$. Suppose we set $x_i = 1$. If the coefficient of t^r is 1 (i.e. if $r \leqslant k$) this indicates that the element x_i can be selected r times. If the term t^r is not present this indicates that the letter x_i cannot be selected r times. More generally, if we want x_i to be chosen a_1 or a_2 or $\cdots$ a_k times ($a_1 < a_2 < \cdots < a_k$), the ith factor should be

$$x_i^{a_1}t^{a_1} + x_i^{a_2}t^{a_2} + \cdots + x_i^{a_k}t^{a_k}. \tag{8}$$

To illustrate: suppose the letter x can only appear an odd number of times. Then its corresponding factor in the generating function is

$$(xt + x^3t^3 + x^5t^5 + x^7t^7 + \cdots).$$

Hence generating functions for all possible types of combinations can be obtained by judiciously choosing the factors. The following examples will help to clarify the procedure.

Example 1 How many 3-combinations can be formed from the letters a, b, c, d if the letter a can be repeated 0, 1, 2, or 3 times and the remaining letters can appear at most once?

The various 3-combinations are listed below:

$$(a, a, a), (a, a, b), (a, a, c), (a, a, d), (a, b, c), (b, c, d,) (a, c, d), (a, b, d).$$

Let us solve the problem using generating functions. The above 3-combinations are generated by

$$(1 + at + a^2t^2 + a^3t^3)(1 + bt)(1 + ct)(1 + dt), \tag{9}$$

since the coefficient of t^3 upon multiplying is $a^3 + a^2b + a^2c + a^2d + abc + bcd + acd + abd$. Hence to find the number of 3-combinations we set $a = b = c = d = 1$ in (9) and find the coefficient of t^3 in

$$(1 + t + t^2 + t^3)(1 + t)^3 = \left(\frac{1 - t^4}{1 - t}\right)(1 + t)^3$$

$$= (1 - t^4)(1 - t)^{-1}(1 + t)^3.$$

This is the same as finding the coefficient of t^3 in $(1 - t)^{-1}(1 + t)^3$ or the coefficient of t^3 in

$$(1 + t + t^2 + t^3 + \cdots)(1 + 3t + 3t^2 + t^3).$$

Hence the coefficient of t^3 is $1 + 3 + 3 + 1 = 8$.

Of course there is no need to use generating functions for such a simple example. However, more complicated problems can be solved by a similar procedure.

Example 2 How many different r-combinations can be formed if the letters $x_1, x_2, \dots, x_k$ can be repeated any number of times while the letters $x_{k+1}, \dots, x_n$ can only be used at most 1 time?

The generating function for this problem is

$$(1 + t + t^2 + \cdots + t^n)^k (1 + t)^{n-k}.$$

Hence we need only find the coefficient of t^r in the last expression or the coefficient of t^r in

$$\left(\frac{1 - t^{n+1}}{1 - t}\right)^k (1 + t)^{n-k} = (1 - t^{n+1})^k (1 - t)^{-k} (1 + t)^{n-k}.$$

Since $r \leqslant n$, $(1 - t^{n+1})^k$ contains powers of t greater than r except for the first term, the problem reduces to finding the coefficient of t^r in $(1 - t)^{-k} (1 + t)^{n-k}$.

A term t^r can appear by choosing t^i from the first factor and t^{r-i} from the second factor ($i = 0, 1, 2, \dots, r$). Hence the coefficient of t^r in the product will be the sum of the product of the coefficient of t^i in the first factor and the coefficient of t^{i-r} in the second factor. The coefficient of t^i in the first factor is $\binom{k + i - 1}{i}$ while the coefficient of t^{r-i} in the second factor is $\binom{n - k}{r - i}$. If $n - k \geqslant r$ the required coefficient is $\sum_{i=0}^{r} \binom{k + i - 1}{i}\binom{n - k}{r - i}$. However if $n - k < r$ the coefficient of t^r is $\sum_{i=r-n+k}^{r} \binom{k + i - 1}{i}\binom{n - k}{r - i}$ since powers of t larger than $n - k$ do not appear in the second factor.

Example 3 The generating function for r-combinations of n distinct objects with unlimited repetitions but with the restriction that each object must appear at least once is

$$(t + t^2 + t^3 + \cdots)^n.$$

This is the same as

$$t^n (1 - t)^{-n} = t^n \sum_{r=0}^{\infty} \binom{n + r - 1}{r} t^r$$

by using (7). Setting $n + r = s$, the last sum reduces to

$$\sum_{s=n}^{\infty} \binom{s-1}{s-n} t^s = \sum_{s=n}^{\infty} \binom{s-1}{n-1} t^s.$$

Hence the number of r-combinations is zero if $r < n$ and $\binom{n-1}{s-1}$ if $r \geqslant n$.

Exercises

1 Write down the generating function for r-combinations formed from the n letters $x_1, \ldots, x_n$ if x_1 is always to appear an even number of times but not more than i times and x_2 is always to appear an odd number of times but not more than j times.

2 Find the number of r-combinations which can be formed from n distinct objects with unlimited repetitions of each object but with each object appearing an even number of times.

3 In how many ways can a total of 100 be obtained upon rolling 50 distinct dice?

4 How many distinct ordered solutions has the equation

$$x + y + z = 50 \quad \text{for } x = 2, 3, \ldots, 50, \quad y = 0, 1, 2, \ldots, 50, \quad z = 5, 6, \ldots, 25?$$

5 In how many different ways can 10 similar flower boxes be painted with 5 different colors provided that each box is painted one solid color?

6 Let $(n_1, n_2, \ldots, n_k)$, where $n_1 + n_2 + \cdots + n_k = n$, denote that there are n_1 things of the first kind, n_2 things of a second kind, ..., n_k things of a k-th kind. For example, $a, a, a, a, b, b, b, c, c, d$ would be specified by (4, 3, 2,1), $aaaa$ by (4) and a, b, c by (1, 1, 1). For convenience we denote symbols such as (1, 1, 1) by 1^3, (2, 2, 3, 3) by $2^2 3^2$ and (4, 1, 1) by 41^2 etc. Show that the generating function for r-combinations of n objects of specification pl^9 is

$$(1 + t + t^2 + \cdots + t^p)(1 + t)^9.$$

Hence find the number of r-combinations.

7 Find the number of r-combinations of n objects of specification s^m.

8 Find the number of r-combinations of n objects of specification $2^m 1^{n-2m}$.

2 Generating functions for permutations

It is impossible to find a generating function, similar to equations 1.3, 1.6, and 1.7 for combinations, in which the sums of all possible permutations appear as coefficients. This is because $x_1 x_2$ and $x_2 x_1$ are indistinguishable under ordinary multiplication. Nevertheless it is still possible to find enumerators whose coefficients are the total *number* of permutations.

In order to find these enumerators we use the fact that r-permutations can be formed in 2 steps. First we find all possible r-combinations and then we permute each r-combination in all possible ways. All possible r-combinations are exhibited as coefficients of t^r in a corresponding combinatorial generating function. For example, the r-combinations of the 3 letters x_1, x_2, x_3 appear as coefficients of t^r ($r = 1, 2, 3$) in equation 1.3. Each r-combination gives rise to $r!$ permutations. If we multiply the coefficient of t^r by $r!$ and divide t^r by $r!$ nothing is changed. We illustrate this process by rewriting equation 1.3 as follows:

$$1 + 1!\,(x_1 + x_2 + x_3)\frac{t}{1!} + 2!\,(x_1x_2 + x_1x_3 + x_2x_3)\frac{t^2}{2!} + 3!\,(x_1x_2x_3)\frac{t^3}{3!}. \tag{1}$$

Since each r-combination gives rise to the same number, namely $r!$ of permutations we need only multiply $r!$ by the number of r-combinations. This is done automatically if we set $x_1 = x_2 = x_3 = 1$ in (1) and look at the coefficient of $\frac{t^r}{r!}$. (1) becomes

$$1 + 3 \cdot 1!\frac{t}{1!} + 2!\,3\,\frac{t^2}{2!} + 3!\,1\,\frac{t^3}{3!}. \tag{1'}$$

In other words, $(1 + t)^3$ on expansion gives ${}_3P_r$ as the coefficient of $\frac{t^r}{r!}$ ($r = 1, 2, 3$).

For n unlike things, it follows in a similar manner from equation 1.4, that

$$(1 + t)^n = \sum_{r=0}^{n} {}_nP_r \frac{t^r}{r!}. \tag{2}$$

Suppose an element may appear 0, 1, 2, ..., k times. For example, suppose we have to find the r-permutations ($r = 1, 2, ..., 5$) which can be formed using the letters a, a, a, b, c. Proceeding as before we first find the symbolic combinatorial generating function, which is

$$(1 + at + a^2t^2 + a^3t^3)(1 + bt)(1 + ct) = 1 + 1!\,(a + b + c)\frac{t}{1!} + 2!\,(bc + a^2 + ab + ac)\frac{t^2}{2!} + 3!\,(a^3 + abc + a^2b + a^2c)\frac{t^3}{3!} + 4!\,(a^2bc + a^3b + a^3c)\frac{t^4}{4!} + 5!\,(a^3bc)\frac{t^5}{5!}. \tag{3}$$

If we set $a = b = c = 1$, the coefficient of $\frac{t^r}{r!}$ $(r = 2, 3, 4, 5)$ is not the number of r-permutations. What has gone wrong? The answer is simple. Not every r-combination gives rise to the same number of permutations. For example, examining the 2-combinations we see that bc gives rise to two permutations bc and cb while aa gives rise to only one. Moreover, the 3-combination a^3 gives rise to $\frac{3!}{3!}$ permutations, abc to $3!$ permutations, while a^2b and a^2c give rise to $\frac{3!}{2!}$ permutations. The reason for these differences is that every time a letter is repeated, say k times, in an r-combination the number of r-permutations is $\frac{r!}{k!}$ (cf. theorem 1.3.2).

Once we realize what the difficulty is, it is easy to correct the trouble. Each time a letter, say a, is repeated p times in a combination, that is, a^p is chosen from its factor in the generating function we must divide by $p!$ This will be automatically taken care of, by the properties of multiplication, if we write $\frac{a^p}{p!}$ as the coefficient of t^p in the corresponding factor of the generating function.

$$\left(1 + a\frac{t}{1!} + a^2\frac{t}{2!} + a^3\frac{t}{3!}\right)\left(1 + b\frac{t}{1!}\right)\left(1 + c\frac{t}{1!}\right)$$

$$= 1 + 1!\,(a + b + c)\frac{t}{1!} + 2!\left(bc + \frac{a^2}{2!} + ab + ac\right)\frac{t^2}{2!}$$

$$+ 3!\left(\frac{a^3}{3!} + abc + \frac{a^2b}{2!\,1!} + \frac{a^2c}{2!\,1!}\right)\frac{t^3}{3!}$$

$$+ 4!\left(\frac{a^2bc}{2!\,1!\,1!} + \frac{a^3b}{3!\,1!} + \frac{a^3c}{3!\,1!}\right)\frac{t^4}{4!} + 5!\left(\frac{a^3bc}{3!\,1!\,1!}\right)\frac{t^5}{5!}. \qquad (3')$$

If we set $a = b = c = 1$, the coefficient of $\frac{t^r}{r!}$ is the number of r-premutations since each r-combination gives rise to the correct number of permutations. Note that each factor of the generating function is in reality a permutation generating function for the corresponding letter.

The reasoning in the general case is similar. We modify the corresponding combinatorial generating function by replacing t^k by $t^k/k!$ in each factor in which it appears. This allows for k repetitions of the corresponding letter whenever it appears in an r-combination. The r-combination would automatically be divided by $k!$ upon multiplying out the factors of the generating function. Hence, if the ith of n letters is to appear $a_1(i), a_2(i), \ldots, a_k(i)$ times $(i = 1, 2, \ldots, n)$, the number of r-permutations is the coefficient of $t^r/r!$ in the product of the permutation generating functions for the individual letters

$$\prod_{i=1}^{n}\left(\frac{t^{a_1(i)}}{a_1(i)!}+\frac{t^{a_2(i)}}{a_2(i)!}+\cdots+\frac{t^{a_{k_i}(i)}}{a_{k_i}(i)!}\right)$$

(cf. equation 1.8 with $x_i = 1$). This general formulation should become clearer upon study of the following examples.

Example 1 The generating function for r-permutations of n different objects with no restriction on the number of times any object may appear is

$$\left(1+t+\frac{t^2}{2!}+\frac{t^3}{3!}+\cdots\right)^n. \tag{4}$$

To expand (4) in powers of t we need to introduce a new function $E(x)$, known as the exponential function, which is defined by

$$E(x) = 1+\frac{x}{1!}+\frac{x^2}{2!}+\cdots=\sum_{r=0}^{\infty}\frac{x^r}{r!} \tag{5}$$

for all real values of x. Since

$$\frac{1}{2!} = 0.5000000,\ \frac{1}{3!} = 0.1666667,\ \frac{1}{4!} = 0.041667,\ \frac{1}{5!} = 0.0083333,$$

$$\frac{1}{6!} = 0.0013889,\ \frac{1}{7!} = 0.0001984,\ \frac{1}{8!} = 0.0000248,\ \frac{1}{9!} = 0.0000028,$$

$$\frac{1}{10!} = 0.0000003;\ \sum_{r=0}^{10}\frac{1}{r!} = 2.7182819\ldots.$$

It can be shown that the series in (5) converges for all x and that $E(1) = 2.71828\ldots$ is an irrational number called e.

We now derive a few properties of this function*.

$$E(x)\,E(y) = \left(\sum_0^\infty \frac{x^r}{r!}\right)\left(\sum_0^\infty \frac{y^r}{r!}\right) = 1 + \frac{x+y}{1!} + \cdots + \frac{x^r}{r!} + \frac{x^{r-1}y}{(r-1)!\,1!}$$

$$+ \frac{x^{r-2}y^2}{(r-2)!\,2!} + \cdots + \frac{x^{r-i}y^i}{(r-i)!\,i!} + \cdots + \frac{y^r}{r!} + \cdots$$

$$= 1 + \frac{x+y}{1!} + \cdots + \frac{1}{r!}\left(x^r + \binom{r}{1}x^{r-1}y + \binom{r}{2}x^{r-2}y^2 + \cdots\right.$$

$$\left. + \binom{r}{i}x^{r-i}y^i + \cdots + y^r\right) + \cdots .$$

By the binomial theorem, $(x^r + \binom{r}{1}x^{r-1}\,y + \cdots + y^r) = (x+y)^r$. Hence the last sum reduces to

$$\sum_{r=0}^{\infty} \frac{(x+y)^r}{r!} = E\,(x+y);$$

the last equality following from (5). Therefore

$$E\,(x+y) = E(x)\,E(y). \tag{6}$$

Setting $x = y$ in (6) yields $E\,(2x) = [E\,(x)]^2$. Hence $E\,(3x) = E\,(2x)\,E\,(x) = [E\,(x)]^3$ and it follows that

$$E\,(mx) = [E(x)]^m \tag{7}$$

for any integer $m > 0$. Replacing m and x by n and $1/n$ respectively, in (7) yields $E(1) = [E(1/n)]^n$ or

$$E\left(\frac{1}{n}\right) = [E(1)]^{1/n}. \tag{8}$$

The substitution $x = 1/n$ in (7) gives

$$E\,(m/n) = [E\,(1/n)]^m = [E(1)]^{m/n} = e^{m/n}$$

from (8). Therefore $E(x) = e^x$ for any rational number $x > 0$. It can be shown that this last result is valid for every real number $x > 0$. From (5),

---

* The following results for a series of positive numbers which converge to s are used: if the terms are rearranged and added the resulting sum is s. If the terms are grouped by the insertion of parentheses and a new series is formed, the ith term being the sum of the terms in the ith group, the new series still converges to s.

$E(0) = 1$. Setting $y = -x$ in (6), $1 = E(x)\,E(-x)$ or

$$E(-x) = \frac{1}{e^x} = e^{-x}. \tag{9}$$

Therefore the function $E(x)$ just represents the number e raised to the xth power.

If $x > 0$, it is evident from (5) that e^x increases as x becomes larger. If $x < 0$, say $x = -y$ with $y > 0$, $e^{-y} = \dfrac{1}{e^y}$ from (9). Hence as x becomes more negative y and hence e^y increases which means that $\dfrac{1}{e^y} = e^x$ decreases. Therefore the graphs of $y = e^x$ and $y = e^{-x}$ have the appearance shown below.

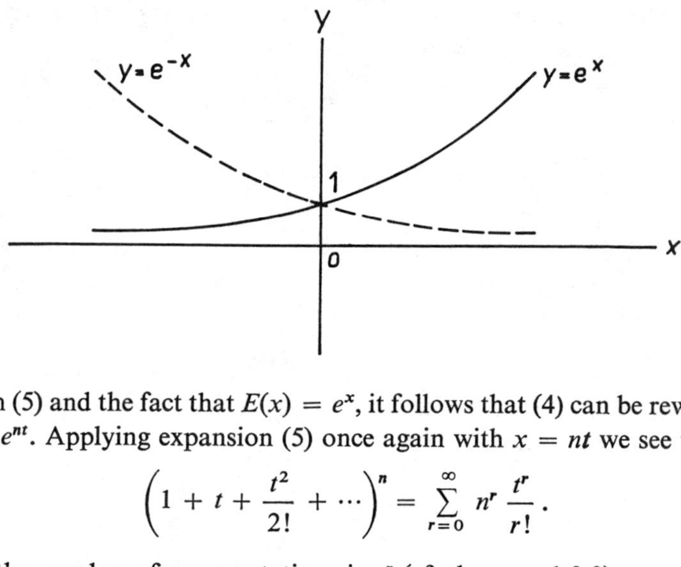

From (5) and the fact that $E(x) = e^x$, it follows that (4) can be rewritten as $(e^t)^n = e^{nt}$. Applying expansion (5) once again with $x = nt$ we see that

$$\left(1 + t + \frac{t^2}{2!} + \cdots\right)^n = \sum_{r=0}^{\infty} n^r \frac{t^r}{r!}.$$

Hence the number of r-permutations is n^r (cf. theorem 1.3.3).

Example 2 The generating function for r-permutations of n objects with the specification of k_1 like things of 1 kind, k_2 like things of another kind, ..., k_s like things of an s-th kind ($k_1 + k_2 + \cdots + k_s = n$) is

$$\left(1 + t + \frac{t^2}{2!} + \cdots + \frac{t^{k_1}}{k_1!}\right)\left(1 + t + \frac{t^2}{2!} + \cdots + \frac{t^{k_2}}{k_2!}\right)$$

$$\cdots\left(1 + t + \frac{t^2}{2!} + \cdots + \frac{t^{k_s}}{k_s!}\right).$$

The reader can verify that the coefficient of $\dfrac{t^n}{n!}$ is $\dfrac{n!}{k_1!\,k_2!\cdots k_s!}$ which agrees with theorem 1.3.2. This result generalizes theorem 1.3.2 since we can find the number of r-permutations by finding the coefficient of $t^r/r!$ For instance, the answer to example 1.4.4 is the coefficient of $t^5/5!$ in

$$\left(1+t+\frac{t^2}{2!}+\cdots+\frac{t^5}{5!}\right)\left(1+t+\frac{t^2}{2!}+\frac{t^3}{3!}\right)^2\left(1+t+\frac{t^2}{2!}\right)^2 \times$$
$$\times\ (1+t)\left(1+t+\frac{t^2}{2!}+\frac{t^3}{3!}+\frac{t^4}{4!}\right).$$

Neglecting powers higher than t^5 we have

$$\left(1+t+\frac{t^2}{2!}+\cdots+\frac{t^5}{5!}\right)\left(1+t+\frac{t^2}{2!}+\cdots+\frac{t^4}{4!}\right)$$
$$=1+2\left(\frac{t}{1!}\right)+4\left(\frac{t^2}{2!}\right)+8\left(\frac{t^3}{3!}\right)+16\left(\frac{t^4}{4!}\right)+31\left(\frac{t^5}{5!}\right)+\cdots \tag{10}$$

$$\left(1+t+\frac{t^2}{2!}+\frac{t^3}{3!}\right)^2\left(1+t+\frac{t^2}{2!}\right)^2$$
$$=1+4\left(\frac{t}{1!}\right)+16\left(\frac{t^2}{2!}\right)+62\left(\frac{t^3}{3!}\right)+228\left(\frac{t^4}{4!}\right)+780\left(\frac{t^5}{5!}\right)+\cdots \tag{11}$$

We leave it to the reader to complete the verification that the coefficient of $t^5/5!$ is 13,431* (cf. example 1.4.4).

The coefficient of $t^r/r!$ in (10) is the number of r-permutations which can be formed from 5 a's and 4b's. The coefficient of $t^r/r!$ in (11) is the number of r-permutations which can be formed from the letters $a, a, a, b, b, b, c, c, d, d$.

Exercises

1 Find the number of r-permutations of n different objects if each object must appear at least once.

2 Write down the generating function for permutations with unlimited repetitions and with each object appearing an odd number of times.

---

* This is just the sum of the coefficients of t^4 and t^5 divided by 5! in the product of (8) and (9). These are 6260 and 7171.

3 What is the generating function for r-permutations of n objects of specification (3^p2^q)?

4 a) Show that the number of r-permutations with repetitions of n different things such that no two consecutive things are alike is $n(n-1)^{r-1}$.

b) Verify that the generating function for this sequence is

$$\frac{nx}{1 - x(n-1)}.$$

3 Distribution of balls in cells

The technique of generating functions is useful in finding the number of ways of assigning balls to boxes or cells. This seemingly abstract problem is actually representative of a large number of familar situations, as shown by the following examples.

(a) The outcomes of a throw with $r = 9$ dice correspond to placing 9 balls into $n = 6$ cells.

(b) The tossing of r coins corresponds to placing r balls in $n = 2$ cells which represent heads and tails.

(c) The possible configurations of the birthdays of r people can be interpreted as the number of ways of placing r balls into $n = 365$ cells.

(d) Suppose a group of r people is classified according to height and weight. The classes play the role of cells and the people that of balls.

(e) Suppose a small boy is collecting pictures placed in boxes of cereal. The different kinds of pictures represent the cells; the pictures collected represent the balls.

(f) In firing at n targets, the targets correspond to cells and the hits to balls.

THEOREM 1 *The number of ways of putting r like objects into n different cells is* $\binom{n+r-1}{r}$.

Proof Consider the product

$$(1 + x_1t + x_1^2t^2 + \cdots + x_1^rt^r)(1 + x_2t + x_2^2t^2 + \cdots + x_2^rt^r)$$
$$\cdots (1 + x_nt + x_n^2t^2 + \cdots + x_n^rt^r) = 1 + h_1t + h_2t^2 + \cdots + h_rt^r + \cdots \quad (1)$$

where h_i $(i = 1, \ldots, r)$ is the sum of all homogeneous products of degree r formed from the n letters $x_1, \ldots, x_n$. This equality was proven in section 1 (see equations 1.5 and 1.6).

We interpret $x_1^{i_1} x_2^{i_2} \cdots x_n^{i_n}$ to mean that there are i_1 balls in cell 1, i_2 balls in cell 2, ..., and i_n balls in cell n, where $i_1 + i_2 + \cdots + i_n$ equals the total number of balls. In other words, the letter x_k represents cell k and the power to which it is raised, i_k, represents the number of balls in cell k. Under this interpretation, a homogeneous product of degree r uniquely represents a distribution of r balls in n cells and vice versa. Hence h_r (1, 1, ..., 1) is the number of ways r balls can be distributed in n cells. Therefore the number of ways is given by the coefficient of t^r in the expansion of $(1 + t + \cdots + t^r)^n$. However the coefficient of t^r in this last expression is the same as the coefficient of t^r in $(1 + t + \cdots + t^r + t^{r+1} + \cdots)^n = \left(\dfrac{1}{1 - t}\right)^n$ since higher order terms have no effect on lower order terms. The coefficient of t^r in $\left(\dfrac{1}{1 - t}\right)^n = (1 - t)^{-n}$ is $\dbinom{n + r - 1}{r}$.

Of course this result could be derived directly by using theorem 4.2 and homogeneous products. The advantage of the above method is that it can be generalized, as we shall see below.

Example 1

a) The number of distinguishable outcomes upon rolling 9 similar dice is $\dbinom{14}{9}$.

b) Five similar books can be arranged on three shelves in $\dfrac{7!}{5!\,2!}$ ways.

THEOREM 2 *The number of ways of putting r similar balls into n different cells so that no cell is empty is* $\dbinom{r - 1}{n - 1}$.

Proof Since no cell is empty x_1^0 does not appear and so we consider

$$(x_1 t + x_1^2 t^2 + \cdots + x_1^r t^r + \cdots)(x_2 t + x_2^2 t^2 + \cdots + x_2^r t^r + \cdots)$$
$$\cdots (x_n t + \cdots + x_n^r t^r + \cdots) \tag{2}$$

When this last expression is expanded, the coefficient of t^r is the sum of all terms of the form $x_1^{i_1} x_2^{i_2} \cdots x_n^{i_n}$ where $i_1 + \cdots + i_n = r$ and $i_j \neq 0$ $(j = 1, \ldots, n)$. Each such term uniquely represents a way of placing r balls in n cells so that no cell is empty and conversely. Hence the required number of ways is the number of such terms which can be obtained by setting $x_1 = x_2 = \cdots$

$= x_n = 1$ and finding the coefficient of t^r. The coefficient of t^r in

$$(t + t^2 + \cdots)^n = t^n (1 - t)^{-n}$$

is

$$\binom{n + (r - n) - 1}{r - n} = \binom{r - 1}{r - n} = \binom{r - 1}{n - 1}.$$

Note that each factor appearing in (1) and (2) is the enumerator for a single cell. The occupancy for each cell is specified independently. Hence if cell i contains at least a balls and at most b the enumerator is

$$x_i^a t^a + x_i^{a+1} t_i^{a+1} + \cdots + x_i^b t^b.$$

THEOREM 3 *The number of ways of placing r different balls in n different cells such that cell k contains i_k balls ($k = 1, 2, \ldots, n$) and the order of the balls in each cell is immaterial is* $\dfrac{r!}{i_1!\, i_2! \cdots i_n!}$.

Proof Again we let $x_1^{i_1} x_2^{i_2} \cdots x_n^{i_n}$ denote the fact that cell k contains i_k balls ($k = 1, 2, \ldots, n$). These terms appear in the expansion of $(x_1 + x_2 + \cdots + x_n)^r$. The coefficient of $x_1^{i_1} x_2^{i_2} \cdots x_n^{i_n}$ is just the number of ways of dividing the r factors into n groups containing $i_1, i_2, \ldots, i_n$ factors. However each such division corresponds to a way of placing the r balls in n cells and vice versa. Hence the required number of ways is just the coefficient of $x_1^{i_1} \cdots x_n^{i_n}$ in $(x_1 + x_2 + \cdots + x_n)^r$ which is $\dfrac{r!}{i_1!\, i_2! \cdots i_n!}$ by the multinomial theorem (2.4.1).

Example 2 In how many ways can 7 different balls be placed in 4 different boxes if the first box contains 2 balls and the order of the balls in a box is immaterial?

By the above theorem, the number of ways is

$$\sum_{i_2+i_3+i_4=5} \frac{7!}{2!\, i_2!\, i_3!\, i_4!}$$

where the summation is over all ordered 3-partitions of 5. This last sum can be rewritten in the following form:

$$\frac{7!}{2!\, 5!} \left(\sum_{i_2+i_3+i_4=5} \frac{5!}{i_2!\, i_3!\, i_4!} \right).$$

The sum appearing in the parentheses can now be evaluated by the corollary to theorem 2.4.1 yielding $\dfrac{7!}{2!\,5!}\,(3^5)$.

The above result as well as the next theorem can easily be established without using generating functions. However the method presented below can be generalized to more complicated problems.

THEOREM 4 *The number of ways of placing r different balls into n different cells, when the order of the balls in each cell is immaterial, is n^r.*

Proof Again we denote the fact that cell k contains i_k balls ($k = 1, \ldots, n$) by $x_1^{i_1} x_2^{i_2} \cdots x_n^{i_n}$. Consider the following product:

$$\left(1 + x_1 t + x_1^2 \frac{t^2}{2!} + \cdots + x_1^r \frac{t^r}{r!}\right)\left(1 + x_2 t + x_2^2 \frac{t^2}{2!} + \cdots + x_2^r \frac{t^r}{r!}\right) \cdots \left(1 + x_n t + x_n^2 \frac{t^2}{2!} + \cdots\right). \tag{3}$$

The coefficient of t^r will consist of a sum of terms of the form $\dfrac{1}{i_1!\, i_2! \cdots i_n!}$ $(x_1^{i_1} x_2^{i_2} \cdots x_n^{i_n})$ where $i_1 + i_2 + \cdots + i_n = r$; the reasoning is similar to that used in theorem 1. Hence the coefficient of $t^r/r!$ will be

$$\sum \frac{r!}{i_1!\, i_2! \cdots i_n!}\, x_1^{i_1} x_2^{i_2} \cdots x_n^{i_n}$$

(where the summation is over all ordered n-partitions of r) or $(x_1 + x_2 + \cdots + x_n)^r$. Now $\dfrac{r!}{i_1!\, i_2! \cdots i_n!}$ is the number of ways of distributing r balls in n cells so that cell k contains i_k balls (by the last theorem) while $x_1^{i_1} \cdots x_n^{i_n}$ describes this particular distribution. Setting $x_1 = x_2 = \cdots = x_n = 1$ in (3), we see that the number of ways is just the coefficient of $t^r/r!$ in $\left(1 + t + \dfrac{t^2}{2!} + \cdots + \dfrac{t^r}{r!}\right)^n$. This is the same as the coefficient of $t^r/r!$ in

$$\left(1 + t + \frac{t^2}{2!} + \cdots\right)^n = e^{nt}.$$

This coefficient is n^r as required.

Note that each factor is the generating function for the occupancy of each cell. The occupancy of each cell is specified independently of the remaining cells. Hence if cell i must contain at least a balls and at most b balls its enumerator is

$$x_i^a \frac{t^a}{a!} + x_i^{a+1} \frac{t^{a+1}}{(a+1)!} + \cdots + x_i^b \frac{t^b}{b!}.$$

We illustrate this procedure in the following theorem.

THEOREM 5 *Suppose that the order of balls within each cell does not count. The number of ways of putting r different balls in n different cells such that m cells are not empty while the remaining are empty is*

$$\binom{n}{m} \sum_{j=0}^{m} \binom{m}{j} (-1)^j (m-j)^r. \tag{4}$$

Proof Suppose that the first m cells are not empty. Then the symbolic generating function is

$$\left(x_1 t + x_1^2 \frac{t^2}{2!} + \cdots\right)\left(x_2 t + x_2^2 \frac{t^2}{2!} + \cdots\right) \cdots \left(x_m t + x_m^2 \frac{t^2}{2!} + \cdots\right). \tag{5}$$

The number of ways that this can be done is the coefficient of $t^r/r!$ in the above expansion when we set $x_1 = x_2 = \cdots = x_m = 1$. For $x_i = 1$, (5) reduces to $(e^t - 1)^m$. The same generating function is obtained no matter which m cells are chosen. Since m cells can be chosen from n cells in $\binom{n}{m}$ ways, the required generating function is $\binom{n}{m}(e^t - 1)$. Expanding in powers of t^r yields

$$\begin{aligned}\binom{n}{m}(e^t - 1)^m &= \binom{n}{m} \sum_{j=0}^{m} \binom{m}{j} (-1)^j e^{(m-j)t} \\ &= \binom{n}{m} \sum_{r=0}^{\infty} \left[\sum_{j=0}^{m} (-1)^j \binom{m}{j} (m-j)^r\right] \frac{t^r}{r!}.\end{aligned}$$

Hence we see that the coefficient of $t^r/r!$ is just (4).

(4) is frequently written in terms of Stirling's number of the second kind, $S(r, n)$ defined by

$$S(r, n) = \frac{1}{n!} \sum_{j=0}^{n} (-1)^j \binom{n}{j} (n-j)^r. \tag{6}$$

Using (6), (4) becomes

$$n(n-1) \cdots (n-m+1)\, S(r, m). \tag{4'}$$

COROLLARY *Suppose the order of balls in a particular cell is immaterial. The number of ways of placing r different balls into n different cells, with no cell empty is*

$$\sum_{j=0}^{n} \binom{n}{j} (-1)^j (n-j)^r = n!\, S(r, n).$$

The following problem is a practical application of theorem 5. Find the number of ways of painting a 10 room house using at most 3 different colors. The colors are to be selected from 20 available colors and each room is to be painted a single color.

The $n = 20$ different colors can be represented by 20 different cells while the $r = 10$ rooms play the role of 10 different balls. Placing a ball into a cell corresponds to painting the corresponding room a particular color. Hence the solution of the problem is equivalent to finding the number of ways of placing 10 different balls into 20 different cells such that 1 or 2 or 3 cells are not empty while the remaining are empty. By (4′) the answer is

$$20 + 20 \cdot 19\, S(10, 2) + 20 \cdot 19 \cdot 18\, S(10, 3).$$

Example 3 To further illustrate this method we will rework example 2. The enumerating function is

$$\frac{t^2}{2!}\left(1 + \frac{t}{1!} + \frac{t^2}{2!} + \cdots + \frac{t^5}{5!}\right)^3.$$

The answer is the coefficient of $t^7/7!$ in the above expression or in $\frac{t^2}{2!}\left(1 + \frac{t}{1!} + \frac{t^2}{2!} + \cdots\right)^3 = \frac{t^2}{2!} e^{3t}$. This is $3^5 \frac{7!}{2!\,5!}$.

The following example illustrates the importance, in some cases, of the order of balls in a cell. We shall call such cells *ordered cells*.

Example 4 In how many ways can a lady without any thumb arrange all her 5 unlike rings on her left hand?

We assume all the rings must be used. Here we have $r = 5$ rings (balls) to be placed on $n = 4$ fingers (cells) where the order of the rings on a finger counts. We first assume that the rings are all alike. By theorem 1, the number of ways they can be placed on her fingers is $\binom{8}{5}$. If the rings are all different, each one of these placements would give rise to 5! different arrangements corresponding to the permutations of the rings. Hence the number of ways is

$$5!\binom{8}{5} = 6720.$$

THEOREM 6 *The number of ways of putting r balls of specification** $(r_1, r_2, \ldots, r_k)$ *where* $(r_1 + r_2 + \cdots + r_k = r)$ *into n different ordered cells is the product of* $\dfrac{r!}{r_1! \cdots r_k!}$ *and the number of ways of putting r like balls into n different unordered cells.*

Proof First distribute the balls as if order in the cells is not important. The number, c, of ways this may be done is just the allowable number of ways that r like balls may be distributed into n unordered cells.

Fix any one of the c distributions. Suppose cell j contains i_j balls $j = 1, 2, \ldots, n$. Since order in the cells is important we must see how many ways this distribution may be ordered. Place the r balls in a row. No matter what the order of the balls, the first i_1 balls belong to cell 1, the next i_2 balls to cell 2, ..., and the last i_n balls to cell n. The order of the balls from left to right is the order in which they appear in the cell. Hence each permutation of the r balls corresponds uniquely to a way of ordering the distributions and vice versa. By theorem 1.3.2 this may be done in $\dfrac{r!}{r_1! \cdots r_k!}$ ways. This number is independent of $i_1, i_2, \ldots, i_n$ and so is independent of the particular distribution chosen. Therefore the total number of distributions is $c\,\dfrac{r!}{r_1! \cdots r_k!}$.

COROLLARY 1 *The number of ways of placing r different balls into n different ordered cells with no restriction on the number in any cell is*

$$r!\binom{n+r-1}{r} = n(n+1)\cdots(n+r-1).$$

COROLLARY 2 *The number of ways of placing r different balls into n different ordered cells so that no cell is empty is*

$$r!\binom{r-1}{n-1}.$$

---

* We use the notation of exercise 1.5.

The proofs of these corollaries follows directly from the theorem by applying theorems 1 and 2 respectively.

Finally let us suppose the cells are alike.

THEOREM 7 *The number of ways of putting r different balls into n similar cells* $(r \geqslant n)$, *with no cell empty is* $S(r, n)$.

Proof Let x be the number of ways this can be done. If we consider the cells to be different, each way yields $n!$ different distributions. Hence the number of ways of placing r distinct balls into n distinct cells with no cell empty is $n!\,x$. By the corollary to theorem 5 this number is $n!\,S(r, n)$. Therefore $n!\,x = n!\,S(r, n)$ which yields the result.

Example 5 In how many ways can the number $2 \cdot 3 \cdot 5 \cdot 7 \cdot 11 \cdot 13 \cdot 17 \cdot 19$ be factored into 3 different numbers?

There are $r = 8$ prime factors of the given number. These play the role of balls. The $n = 3$ different factors correspond to cells. Hence by the above theorem the number of ways is

$$S(8, 3) = \frac{1}{6} \sum_{j=0}^{3} (-1)^j \binom{3}{j} (3 - j)^8 .$$

THEOREM 8 *The number of ways of placing r different balls into n like cells without restrictions is*

$$S(r, 1) + S(r, 2) + \cdots + S(r, n). \tag{7}$$

Proof Any distribution of balls in cells has 0 or 1 or 2 or $\cdots$ or $n - 1$ empty cells. The order in which the empty cells appear does not count; only the number of empty cells is pertinent. If $n - i$ cells are empty ($i = 1, 2, \ldots, n$) there are i cells which are not empty and the r balls can be distributed into these i cells in $S(r, i)$ ways by theorem 7. Hence $S(r, i)$ is the number of distributions having $n - i$ empty cells. Therefore by the corollary to theorem 1.2.2 the result is given by (7).

Let us denote the number of unordered k-partitions of r, where each part is a positive integer, by $p_k(r)$. For example, $p_2(5) = 2$ and $p_3(8) = 5$.

THEOREM 9 *The number of ways of putting r similar balls into n similar cells, with no cell empty, is* $p_n(r)$.

The proof of the above result is obvious. By the same reasoning used in the proof of theorem 8 we can prove the following result by using theorem 9.

THEOREM 10 *The number of ways of putting r similar balls into n like cells with no restrictions is*

$$p_1(r) + p_2(r) + \cdots + p_n(r).$$

Unfortunately there is no simple analytic formula for $p_n(r)$. However there are tables of partitions.*

The reader interested in pursuing this topic further should consult P. A. MacMahon, *Combinatory Analysis*, Vol. 1, Cambridge (1915).

Exercises

1 a) Show that the generating function of any cell is $(x + n - 1)^r$ when r distinct objects are distributed over n distinct cells.

b) Generalize by showing that the enumerator for the occupancy of k cells is

$$(x_1 + x_2 + \cdots + x_k + n - k)^r.$$

2 If there are n distinct cells and each cell must contain at least 5 balls show that the generating function for the total number of ways is

$$\left(e^t - 1 - t - \cdots - \frac{t^{s-1}}{(s-1)!}\right)^n.$$

3 If n different cells may contain no less than k of r similar objects, show that the enumerator is $t^{nk}(1 - t)^{-n}$.

4 a) Show directly, by using theorems 1 and 4, that the number of ways of putting r objects of specification $(1^{r_1}2^{r_2}\cdots)$ into n different cells is

$$n^{r_1}\binom{n+1}{2}^{r_2}\binom{n+2}{3}^{r_3}\cdots.$$

b) Let $G(t_1; x_1, x_2, \cdots, x_n) = 1 + h_1t_1 + \cdots + h_kt_1^k + \cdots$ be the generating function for objects of the first kind, where $h_k = h_k(x_1, x_2, \cdots, x_n)$ is the homogeneous product of degree k and represents the sum of the ways of placing k similar balls in n different cells. Similarly, let

$$H(t_2; x_1, x_2, \cdots, x_n) = 1 + h_1t_2 + \cdots + h_kt_2^k + \cdots$$

be the generating function for objects of the second kind. Then verify that the enumerator for objects of both kinds is

$$E(t_1, t_2; x_1 \cdots x_n) = GH = \sum\sum h_ih_jt_1^it_2^j.$$

c) Use part (b) to show that the enumerator for occupancy of n different cells by objects of specification $(1^{r_1}2^{r_2}\cdots)$ is $h_1^{r_1}h_2^{r_2}\cdots$. Hence deduce the result given in (a).

---

* H. Gupta, *Tables of Partitions*, Madras (1939).

5 a) Show that the enumerator for the occupancy of a single cell by k different types of objects is $G(x; t_1, t_2, \cdots, t_k) = \prod_{i=1}^{k} (1 - xt_i) = 1 + h_1(t_1, t_2, \cdots, t_k)\,x + h_2(t_1, t_2, \cdots, t_k)\,x^2 + \cdots + h_r(t_1 \cdots t_k)\,x^r + \cdots$ if we interpret the term $(xt_1)^{r_1}(xt_2)^{r_2} \cdots (xt_k^{r_k})$ as the occupancy of the given cell by $r = r_1 + \cdots + r_k$ objects of wich r_1 are of the first kind, r_2 of the second kind and so on.

b) Show that the enumerator for n unlike cells is

$$\prod_{i=1}^{n} G(x_1, t_1, t_2, \cdots, t_k).$$

c) Show that when no cell is empty and no attention is paid to which cell is occupied the enumerator is

$$(h_1x + h_2x^2 + \cdots)^n.$$

6 Suppose we have $n = n_1 + n_2 + \cdots + n_k$ balls in a box of which n_i are of color i $(i = 1, 2, \cdots, k)$. The balls are withdrawn one by one. The first m_1 balls are placed in a box of color 1, the next m_2 balls in a box of color 2 and so on. Show that the coefficient of $x_1^{m_1}x_2^{m_2} \cdots x_k^{m_k}t^r$ in the expansion of

$$(x_1t + x_2 + x_3 + \cdots + x_k)^{n_1}(x_1 + x_2t + x_3 + \cdots + x_k)^{n_2} \times$$
$$\times (x_1 + x_2 + x_3t + x_4 + \cdots + x_k)^{n_3} \cdots (x_1 + x_2 + \cdots + x_kt)^{n_k}$$

is the number of ways in which r balls match the color of the box containing them.

4 Recursion

Sometimes the answer a_n to a combinatorial problem reduces to finding the solution of a recurrence equation—for example,

$$a_n = 2a_{n-1} + 3a_{n-2} \qquad n = 2, 3, \ldots \tag{1}$$

given that $a_0 = 1$ and $a_1 = 2$. Then from (1),

$$a_2 = 2a_1 + 3a_0 = 7;$$
$$a_3 = 2a_2 + 3a_1 = 14 + 6 = 20;$$
$$a_4 = 2a_3 + 3a_2 = 61.$$

Continuing in this manner, a_n can eventually be found for any integer n. This numerical approach, even for more complicated recurrences, is quite practical especially if a computer is used. However, in some situations, it is desirable to obtain an analytic formula for a_n. Generating functions can be used for this purpose.

Let $f(x) = \sum_{i=0}^{\infty} a_i x^i$ be the generating function of the sequence $a_0, a_1, a_2, \ldots$ and $g(x) = \sum_{i=0}^{\infty} b_i x^i$ be the generating function of the sequence $\{b_i\}$. The b's will be determined later. Then

$$f(x)\, g(x) = \sum_{n=0}^{\infty} c_n x^n \tag{2}$$

where

$$c_n = a_n b_0 + a_{n-1} b_1 + a_{n-2} b_2 + \cdots + a_{n-r} b_r + \cdots + a_0 b_n, \quad n = 0, 1, 2 \ldots \tag{3}$$

Set $c_n = a_n - 2a_{n-1} - 3a_{n-2}$ for $n = 2, 3, \ldots$; then from (3)

$$a_n - 2a_{n-1} - 3a_{n-2} = a_n b_0 + a_{n-1} b_1 + \cdots + a_0 b_n, \quad n = 2, 3, \ldots . \tag{4}$$

Regarding (4) as an identity in the a's and b's it follows that $b_0 = 1, b_1 = -2$, $b_2 = -3$ and $b_i = 0$ for $i \geqslant 3$. By definition of c_n and (1), $c_n = 0$ for $n = 2$, $3 \ldots$ while $c_0 = a_0 b_0 = 1$ and $c_1 = a_0 b_1 + a_1 b_0 = 0$. Hence (2) reduces to

$$f(x)\,(1 - 2x - 3x^2) = 1. \tag{5}$$

Solving (5) for $f(x)$ yields

$$f(x) = \frac{1}{1 - 2x - 3x^2} = \frac{1}{(1 - 3x)(1 + x)} = \frac{3}{4}\left(\frac{1}{1 - 3x}\right) + \frac{1}{4}\left(\frac{1}{1 + x}\right). \tag{6}$$

Since $\dfrac{1}{1 - 3x} = \sum_{i=0}^{\infty} 3^i x^i$ and $\dfrac{1}{1 + x} = \sum_{i=0}^{\infty} (-1)^i x^i$, it follows from (6) upon equating coefficients of x^n that

$$a_n = \tfrac{3}{4}(3)^n + \tfrac{1}{4}(-1)^n.$$

More generally, suppose that the sequence $a_0, a_1, a_2, \ldots$ satisfies the recurrence of the rth order:

$$a_n = \alpha_1 a_{n-1} + \alpha_2 a_{n-2} + \cdots + \alpha_r a_{n-r} \quad n = r, r + 1, r + 2, \ldots \tag{7}$$

given that $a_n = \beta_n$ for $n = 0, 1, 2, \ldots, r - 1$ where the α's and β's are constants. Proceeding as above, define c_n by

$$c_n = a_n - \alpha_1 a_{n-1} - \alpha_2 a_{n-2} - \cdots - \alpha_r a_{n-r} \quad n = r, r + 1, r + 2, \ldots$$

From (3) and the above equation for $n \geqslant r$ it follows that $b_0 = 1$, $b_1 = -\alpha_1$, $\ldots, b_r = -\alpha_r$; and $b_i = 0$ for $i \geqslant r + 1$.

By definition of c_n and (7) $c_n = 0$ for $n \geqslant r$. Using (3) we obtain

$$c_0 = a_0 b_0 = \beta_0,\ c_1 = a_1 b_0 + a_0 b_1 = \beta_1 - \alpha_1 \beta_0,\ \ldots,$$
$$c_{r-1} = \beta_{r-1} - \beta_{r-2}\alpha_1 - \cdots - \beta_0 \alpha_{r-1}.$$

Hence, from (2),

$$f(x) = \frac{c_0 + c_1 x + \cdots + c_{r-1}x^{r-1}}{1 - \alpha_1 x - \alpha_2 x^2 - \cdots - \alpha_r x^r} \equiv \frac{n(x)}{d(x)}. \tag{8}$$

Suppose that the denominator $d(x)$ in (8) can be factored—namely,

$$d(x) = (1 - t_1 x)^{h_1} (1 - t_2 x)^{h_2} \cdots (1 - t_s x)^{h_s}$$

where $h_1 + h_2 + \cdots + h_s = r$. Each t_i^{-1} is a root of the equation $d(x) = 0$ which is repeated h_i times. It can be shown that the rational function $f(x)$ can be expressed in terms of partial fractions

$$f(x) = \sum_{i=1}^{s} \sum_{k=1}^{h_i} \frac{A_{ik}}{(1 - t_i x)^k}, \tag{9}$$

where

$$A_{ik} = \frac{1}{(-t_i)^{h_i-k}(h_i - k)!} \left[\frac{d^{h_i-k}}{dx^{h_i-k}} (1 - t_i x)^{h_i} f(x) \right]_{x=t_i^{-1}}.$$

By the binomial theorem,

$$\frac{A}{(1 - tx)^k} = A\left(1 + (-k)(-tx) + \cdots + \frac{(-k)\cdots(-k+n-1)}{n!}(-tx)^n + \cdots\right).$$

Therefore, comparing coefficients of x^n in (9),

$$a_n = \sum_{i=1}^{s} \sum_{k=1}^{h_i} A_{ik} \binom{n+k-1}{k-1} t_i^n.$$

In particular, if $h_i = 1$ for all i, that is, if there are no repeated roots

$$a_n = \sum_{i=1}^{r} A_i t_i^n \tag{10}$$

where

$$A_i = \lim_{x \to t_i^{-1}} (1 - t_i x) f(x) = n(t_i^{-1})/d'(t_i^{-1})$$
$$= n(t_i^{-1})/(1 - t_1/t_i)(1 - t_2/t_i)\cdots(1 - t_{i-1}/t_i)(1 - t_{i+1}/t_i)\cdots(1 - t_r/t_i). \tag{11}$$

Example 1 In how many ways can a boy climb up a flight of n steps if he can skip at most one step at a time?

Let a_n ($n \geqslant 2$) be the number of ways that the boy can climb the stairs. If the boy steps on the first step the remaining $n - 1$ stairs can be climbed in a_{n-1} ways. On the other hand, if he skips the first step the remaining $n - 2$ steps can be climbed in a_{n-2} ways. Since there are only two possibilities,

$$a_n = a_{n-1} + a_{n-2} \qquad n = 3, 4, \ldots. \tag{12}$$

It is obvious that $a_1 = 1$ and $a_2 = 2$. Note that $a_{n+1} = F(n)$ the Fibonacci numbers introduced in exercise I.4.16. In order to apply the general results derived above, set $a_0 = 1$. This is consistent with (12). From (8)

$$f(x) = \frac{1}{1 - x - x^2}.$$

The roots of $1 - x - x^2$ are $t_1^{-1} = -\dfrac{1 + \sqrt{5}}{2}$ and $t_2^{-1} = -\dfrac{1 - \sqrt{5}}{2}$.

Therefore from (11), $A_1 = \dfrac{1}{2} - \dfrac{\sqrt{5}}{10}$ and $A_2 = \dfrac{1}{2} + \dfrac{\sqrt{5}}{10}$ and so the solution of (12) is, by using (10),

$$a_n = \left(\frac{1}{2} - \frac{\sqrt{5}}{10}\right)\left(\frac{1 - \sqrt{5}}{2}\right)^n + \left(\frac{1}{2} + \frac{\sqrt{5}}{10}\right)\left(\frac{1 + \sqrt{5}}{2}\right)^n \qquad n = 1, 2, 3, \ldots$$

Example 2 Into how many regions is the plane separated by n straight lines such that no two of the lines are parallel and no three of the lines meet at a point?

Let a_n denote the number of regions into which n such lines separate the plane. It can easily be seen that $a_1 = 2$. First, consider $n - 1$ lines in the plane; these separate it into a_{n-1} regions. Then add the nth line. This line must intersect all of the other straight lines because no two lines are parallel. As we move along the nth line from left to right each time we pass through a point of intersection we enter a new region (the line we met is a boundary of the two regions). Since no three of the lines are concurrent it is easy to see that the nth line must pass through n regions. Each time the nth line passes through an existing region (one of the regions determined by the $n - 1$ lines) it divides this region into two regions whose common boundary is the nth line. Hence the net effect of adding the nth line is the creation of n new regions and so there are now $a_{n-1} + n$ regions. This being the total number of regions into which n straight lines separate the plane, it follows that

$$a_n = n + a_{n-1}. \tag{13}$$

Setting $a_0 = 1$, which is consistent with (13), and summing both sides of (13) from 1 to k yields

$$\sum_{n=1}^{k} a_n = \sum_{n=1}^{k} n + \sum_{n=1}^{k} a_{n-1} = \frac{k(k+1)}{2} + \sum_{i=0}^{k-1} a_i = \frac{k(k+1)}{2} + 1 + \sum_{n=1}^{k-1} a_n.$$

Since $\sum_{n=1}^{k} a_n - \sum_{n=1}^{k-1} a_n = a_k$; the last equation implies that

$$a_k = \frac{k(k+1)}{2} + 1 = \frac{k^2 + k + 2}{2} \qquad k = 1, 2, 3, \ldots.$$

Example 3 It is well known that the operations of addition and multiplication are associative, that is, for any real numbers a, b, and c

$$(a + b) + c = a + (b + c), \quad (ab)\,c = a\,(bc).$$

However, there are many operations in mathematics that are nonassociative. For example, suppose we define the operation by

$$a \circ b = \frac{a + b}{2}.$$

Then $1 \circ (2 \circ 3) = \left(1 + \frac{2 + 3}{2}\right)\Big/2 = \frac{7}{4}$ while $(1 \circ 2) \circ 3 = \frac{9}{4}$. Hence $1 \circ (2 \circ 3) \neq (1 \circ 2) \circ 3$. To simplify the typography we shall write $a \circ (b \circ c)$ as $a\,(bc)$ and call it a product. We assume that such products are nonassociative, that is, $(ab)\,c$ is different from $a\,(bc)$. Find the number of interpretations a_n of $x_1 x_2 \cdots x_n$ when parentheses are inserted in all possible ways.

For $n = 3$, we have the possibilities $x_1\,(x_2 x_2)$ and $(x_1 x_2)\,x_3$. If $n = 4$ the various possible products are:

$$x_1\,(x_2\,(x_3 x_4)),\ x_1\,((x_2 x_3)\,x_4),\ (x_1 x_2)\,(x_3 x_4),\ (x_1\,(x_2 x_3))\,x_4,\ ((x_1 x_2)\,x_3)\,x_4.$$

Thus $a_3 = 2$ and $a_4 = 5$. Obviously $a_2 = 1$ and we set $a_1 = 1$ by convention. A product of n letters must be of the form $(x_1 \cdots x_r)\,(x_{r+1} \cdots x_n)$ for $r = 1$ or 2 or $\cdots$ or $(n - 1)$. Where $(x_1 \cdots x_r)$ denotes a product of the first r letters and $(x_{r+1} \cdots x_n)$ is some product of the remaining $n - r$ letters. There are $a_r a_{n-r}$ products of this type since there are a_r products of the form $(x_1 \cdots x_r)$ and for each of these, a_{n-r} products can be formed from the remaining $n - r$ letters. The convention $a_1 = 1$ is consistent with the situations $r = 1$ and $r = n - 1$. Hence

$$a_n = a_1 a_{n-1} + a_2 a_{n-2} + \cdots + a_{n-1} a_1, \quad n \geqslant 2.$$

The above recursion is equivalent to the relation

$$(f(x))^2 = -x + f(x) \tag{14}$$

where the generating function $f(x) = \sum_{i=1}^{\infty} a_i x^i$. Solving (14) as a quadratic equation in $f(x)$ yields

$$f(x) = \frac{1 - \sqrt{1 - 4x}}{2}. \tag{15}$$

Here the minus sign must be used since the series for $f(x)$ has no constant. Expanding the right-hand side of (15) as a power series proves that the coefficient of x_n is

$$a_n = \frac{(2n - 2)!}{n!\,(n - 1)!}, \quad n \geqslant 2.$$

Example 4 Let $N(r, k, m)$ be the number of distinct ways in which r similar balls can be distributed in k different cells allowing at most m balls per cell. Find a recursion relation for $N(r, k, m)$.

If cell 1 contains j balls, where $j = 0$ or 1 or $\cdots$ or m, the remaining $r - j$ balls can be distributed in the other $k - 1$ cells in $N(r - j, k - 1, m)$ ways. Hence

$$N(r, k, m) = \sum_{j=0}^{m} N(r - j, k - 1, m). \tag{16}$$

Note that $N(r, k, 1) = \binom{k}{r}$. Although in general we cannot obtain an analytic solution for (16) we can obtain a numerical solution. To illustrate the procedure, suppose that $m = 3$; then (16) reduces to

$$N(r, k, 3) = N(r, k - 1, 3) + N(r - 1, k - 1, 3) + N(r - 2, k - 1, 3) + N(r - 3, k - 1, 3) \tag{17}$$

By convention $N(0, 0, 3) = 1$ and $N(r, 0, 3) = 0$ for $r > 0$. It is obvious that $N(r, 1, 3) = 1$ for $r = 0, 1, 2, 3$ and $N(r, 1, 3) = 0$ for $r > 3$. More generally, $N(r, k, 3) = 0$ for $r > 3k$ since at most 3 balls can be in each cell. Additional values of $N(r, k, 3)$ are given in the following table:

$k \backslash r$	0	1	2	3	4	5	6	7	8	9
0	1									
1	1	1	1	1						
2	1	2	3	4	3	2	1			
3	1	3	6	10	12	12	10	6	3	1
4	1	4	10	20	31	40	.	.	.	.

From (17) it is not difficult to see that each entry (except $k = 0$) in the table was obtained by taking the sum of the value immediately above and the three entries (foı $r \geqslant 2$) to the left of the one immediately above. For example for $r = 5$ and $k = 3$ we had $12 = 2 + 3 + 4 + 3$.

Exercises

1 Rework example 1 under the assumption that the boy can skip at most two steps at a time.

2 (The Tower of Hanoi Puzzle) There are n circular hoops placed over one of three vertical pegs. The hoops have unequal radii and the largest hoop is at the bottom of the pile on one peg and is covered successively by smaller hoops so that the smallest one is on top. Find a way of transferring the tower of hoops from the peg on which they rest initially to one of the other two pegs under the following rule: The hoops may be moved freely, one at a time, from peg to peg, except that no hoop can ever be placed on top of a smaller hoop. Calculate the number of moves required.

3 Let $a(n, k)$ be the number of regions into which the plane is separated by $n + k$ lines which satisfy the following conditions: (a) exactly k of the lines are parallel to each other (b) no three of the lines are concurrent. Show that

$$a(n, k) = n + 1 + a(n, k - 1)$$

and solve this recurrence for $a(n, k)$.

4 Into how many regions is the interior of a circle divided by the line segments joining n points on the circumference?

5 Let n planes in 3-dimensional space satisfy the following conditions:
(a) No two are parallel. (b) No two lines of intersection are parallel. (c) No four intersect in a point. Into how many regions is the space separated by these planes?

6 Let $N(r, k, m, i)$ be the number of ways in which r similar balls can be placed in k different cells allowing at most m balls per cell if i cells contain the maximum number. Show that

$$N(r, k, m, i) = \binom{k}{i} N(r - mi, k - i, m - 1)$$

where $N(r, k, m)$ is defined in example 4.

7 A derangement is a permutation $a_1, a_2, \ldots, a_n$ of $1, 2, \ldots, n$ such that $a_i \neq i$ for every $i = 1, 2, \cdots, n$. Let d_n be the number of derangements. Show that

$$d_n = (n - 1)\, d_{n-1} + (n - 1)\, d_{n-2}$$

and find an analytic formula for d_n (cf. example 4.2.1).

8 Let $p_k(n)$ be the number of (unordered) partitions of n into k parts (cf. theorem 3.9). Show that

$$p_k(n) = p_k(n - k) + p_{k-1}(n - k) + \cdots + p_1(n - k).$$

9 a) Obtain the generating function $f(x)$ for the sequence $\{a_n\}$ defined in (1), by multiplying (1) by x^n and summing from $n = 2$ to $n = \infty$ and using the fact that $a_0 = 1$ and $a_1 = 2$.

b) Deduce (8) by this method.

10 (Solution of difference equations). Consider

$$a_n + \alpha_1 a_{n-1} + \alpha_2 a_{n-2} + \cdots + \alpha_r a_{n-r} = 0 \quad (\alpha_r \neq 0)$$

where the α_i are real constants. Suppose that $a_i = \beta_i$ for $i = 1, 2, \ldots, r$. Set $a_n = x^n$ in the above equation; it reduces to $x^r + \alpha_1 x^{r-1} + \alpha_2 x^{r-2} + \cdots + \alpha_r = 0$.
Assume that the left-hand side factors into

$$(x - x_1)^{h_1} (x - x_2)^{h_2} \cdots (x - x_s)^{h_s} \quad (h_1 + h_2 + \cdots + h_s = r)$$

where each x_i is a root which is repeated h_i times. Then the solution of the difference equation (or recurrence relation) is

$$u_n = x_1^n (A_{10} + A_{11}x^1 + A_{12}x^2 + \cdots + A_{1h_1-1}x_1^{h_1-1}) + \cdots$$
$$+ x_s^n (A_{s0} + A_{s1} + A_{s2}x^2 + \cdots + A_{sh_s-1}x^{h_s-1}).$$

The r constants A_{ij} are found by setting $u_n = b_n$ for $n = 1, 2, \ldots, r$ and solving the resulting r equations for the A_{ij}. Use this method to solve (1).

Miscellaneous problems

1 Find P_n the number of ways of paying the amount of n cents with five kinds of coins: cents, nickels, dimes, quarters and half-dollars.*

2 Find the number D_n of different dissections of a convex polygon with n sides into $n - 2$ triangles by $n - 3$ diagonals.

---

* Solutions to problems 1—3 can be found in G. Polya, "On picture-writing", *Am. Math. Mon.*, **63** (1956).

3 A *tree* is a connected system of two kinds of objects, lines and points, that contains no closed path (see the diagrams below).

A certain point of the tree in which just one line ends is called the *root* of the tree, the line starting from the root the *trunk*, any point different from the root a *knot*. Compute the number of different trees with n knots.

4 Rework the examples in I.5 by using generating functions.

CHAPTER IV

The principle of inclusion and exclusion

1 More counting principles

IN THIS SECTION we want to count the number of objects having certain prescribed properties. Suppose that there are N objects. We shall use capital letters $A, B, C, \ldots$ with or without subscripts to denote properties of these objects. The number of objects with property A will be denoted by $N(A)$.

Having the property A or B is also a property which will be denoted by $A \cup B$. The word "or" will be used in the following sense: an object has the property A or B if and only if it either has the property A or it has the property B or else it has both the property A and B. The property of having both the property A and the property B will be denoted by AB. The symbol A' will denote the absence of the property A; it is just the property "not A".

Example 1 Let the objects under consideration be chairs. Let Y be the property that a chair is partly yellow and R be the property that a chair is partly red. We write Y = yellow and R = red for short. Then

YR = yellow and red $\qquad Y \cup R$ = yellow or red

Y' = not yellow $\qquad R'$ = not red

$Y'R'$ = not red and not yellow $\qquad Y' \cup R$ = not yellow or red

Suppose there are $N = 5$ chairs colored: red; yellow; red and yellow; green; red, white and blue. Then

$$N(Y) = 2 \qquad N(RY) = 1 \qquad N(R' \cup Y') = 4$$

$$N(R) = 3 \qquad N(RY') = 2 \qquad N(R' \cup Y) = 3$$

$$N(Y') = 3 \qquad N(R'Y) = 1 \qquad N(R \cup Y') = 4$$

$$N(R') = 2 \qquad N(R'Y') = 1 \qquad N(R \cup Y) = 4$$

Generalizing the above notation, we will denote the property "A_1 or A_2 or $\cdots$ or A_n" by $A_1 \cup A_2 \cdots \cup A_n$ and the property "A_1 and A_2 and $\cdots$ and A_n" by $A_1A_2 \cdots A_n$. For instance, using the notation of example 1, let

G = green and B = blue. Then $R \cup Y \cup G$ = yellow or red or green while $R \cup Y \cup B$ = yellow or red or blue; $N(R \cup Y \cup G) = 5$ while $N(R \cup Y \cup B) = 4$. On the other hand, $R\,Y\,B$ = red and yellow and blue and $N(R\,Y\,B) = 0$.

An object has at least one of the properties A_i ($i = 1, 2, \ldots, n$) if and only if it has the property $A_1 \cup A_2 \cup \cdots \cup A_n$. An object possesses all of the properties A_i if and only if it has the property $A_1A_2 \cdots A_n$. Since the same property can be described in different ways, a definition of equality of properties for the objects under consideration is essential. Two properties A and B are *equal*, written $A = B$, if and only if every object possessing property A also possesses property B and conversely. The following lemmas illustrate this definition.

LEMMA 1 $(A')' = A$.

Proof The proof rests upon the fundamental fact of logic that every object either has the property A or it does not have the property A—that is—it has the property A', but it cannot have both.

Suppose an object has the property $(A')'$. This means that the object does not have the property A'. Lacking the property A' it must have the property A.

If the object has the property A then it does not have the property A' and so it has the property $(A')'$. The definition of equality completes the proof.

LEMMA 2 $(A_1 \cup A_2 \cup \cdots \cup A_n)' = A_1'A_2' \cdots A_n'$.

Proof If an object possesses the property $(A_1 \cup A_2 \cup \cdots \cup A_n)'$ it does not have the property $A_1 \cup A_2 \cup \cdots \cup A_n$ and so it does not possess the property A_i for any i. Hence it has the properties A_1' and A_2' and ... and A_n' or in other words the property $A_1'A_2' \cdots A_n'$.

Suppose an object has the property $A_1'A_2' \cdots A_n'$. Then it possesses the properties $A_1', A_2', \ldots, A_n'$ and so cannot have any of the properties A_i ($i = 1, \ldots, n$). Therefore it does not have the property $A_1 \cup A_2 \cup \cdots \cup A_n$ and so it has the property $(A_1 \cup A_2 \cup \cdots \cup A_n)'$.

COROLLARY $(B_1B_2 \cdots B_n)' = B_1' \cup B_2' \cup \cdots \cup B_n'$.

Proof Set $A_i = B_i'$ in the lemma. We have

$$(B_1' \cup B_2' \cup \cdots \cup B_n')' = (B_1')'\,(B_2')' \ldots (B_n')'.$$

From lemma 1, $(B_i')' = B_i$ and so the above equation reduces to

$$(B_1' \cup B_2' \cup \cdots \cup B_n')' = B_1 B_2 \cdots B_n.$$

Using the fact that $C = D$ implies $C' = D'$ in the last equation yields

$$((B_1' \cup B_2' \cup \cdots \cup B_n')')' = (B_1 B_2 \cdots B_n)'.$$

Another application of lemma 1 to the left side gives the result.

Example 2

1. $((A')')' = A'$.
2. $(AB)' = A' \cup B'$. $\qquad (A' \cup B')' = A''B'' = AB$.
3. $(A'B)' = A'' \cup B' = A \cup B'$.
4. $(A' \cup B)' = A''B' = AB'$.

In example 1 we see that

$$N = N(Y) + N(Y') = N(R) + N(R') = N(YR) + N(Y' \cup R')$$
$$= N(RY') + N(R' \cup Y) \cdots$$

This suggests the following result.

THEOREM 1 $N(A') = N - N(A)$.

Proof Every object either has the property A or the property A' but not both. Hence the N objects can be separated into two classes. The first consisting of all objects with the property A, while the second consists of all objects with the property A'. The number of objects in these classes is $N(A)$ and $N(A')$ respectively. Since each object must lie in one of these classes and no object lies in both of these classes, $N(A) + N(A') = N$.

COROLLARY $N(A_1'A_2' \cdots A_n') = N - N(A_1 \cup A_2 \cup \cdots \cup A_n)$.

The proof follows directly from the theorem if we set $A = A_1 \cup A_2 \cup \cdots \cup A_n$ and apply lemma 2.

The following example illustrates another counting principle.

Example 3 Suppose we have 10 red, 5 green, and 6 red and green marbles. Let R be the property that a marble is partly red and G be the property that a marble is partly green; then

$$21 = N(R \cup G) = N(R) + N(G) - N(RG).$$

The reader can verify that a similar result holds in example 1.

THEOREM 2

$$N(A \cup B) = N(A) + N(B) - N(AB). \tag{1}$$

Proof An object has the property $A \cup B$ if and only if it has the property A or the property B or both A and B. If it possesses solely the property A then it will only be counted once in computing the value of the right side of (1)—namely, in $N(A)$. A similar argument holds for objects with property B. An object with both the property A and B will be counted once in $N(A)$, $N(B)$, and $N(AB)$ and so will be counted a total of $1 + 1 - 1 = 1$ times according to the right side of (1). Hence every object with the property $A \cup B$ is counted exactly once in the right side of (1) and only objects with the property $A \cup B$ are counted.

The phrase, *inclusion* and *exclusion*, refers to the process of including everything and then excluding those not required (in the theorem above, the objects repeated twice because they have the property AB). We shall see in the general situation that the above process usually has to be repeated.

COROLLARY

$$N(A'B') = N - N(A) - N(B) + N(AB). \tag{2}$$

The proof of the corollary follows directly from the theorem upon noting that $N - N(A \cup B) = N(A'B')$. Of course (2) can be proved directly using the principle of inclusion and exclusion.

Example 4 How many positive integers less than 1000 are not divisible either by 5 or 7?

Let a positive integer have the property $A(B)$ if and only if it is less than 1000 and divisible by 5(7) respectively. Then* $N(A) = [999/5] = 199$ since the integers with the property A are: $5, 10, 15, \ldots, 5(199) = 995$. Similarly $N(B) = [999/7] = 142$. Integers with the property AB consist of all positive integers $\leqslant 999$ which are multiples of 35, since they are divisible by 7 and 5.

* $[x]$ denotes the greatest integer $\leqslant x$. For instance $[5/2] = 2$, since $2 \leqslant 2\frac{1}{2}$; $[10.9] = 10$, $[6] = 6$, and $[-7 - 1] = -8$.

Hence $N(AB) = [999/35] = 28$. The integers less than 1000 which are divisible neither by 5 nor 7 have the property $A'B'$. By (2), $N(A'B') = 999 - 199 - 142 + 28 = 686$.

THEOREM 3

$$N(A_1 \cup A_2 \cup \cdots \cup A_k) = N(A_1) + \cdots + N(A_k) - N(A_1A_2) - \cdots$$
$$- N(A_{k-1}A_k) + N(A_1A_2A_3) + \cdots + (-1)^{k-1} N(A_1 \cdots A_k)$$
$$= \sum N(A_i) - \sum N(A_iA_j) + \sum N(A_iA_jA_k) + \cdots$$
$$+ (-1)^{r-1} \sum N(A_{i_1}A_{i_2} \cdots A_{i_r}) + \cdots + (-1)^{k-1} N(A_1A_2 \cdots A_k) \tag{3}$$

where the summation is over all r-combinations of the integers $1, 2, \ldots, k$ *for* $r = 1, 2, \ldots, k$.

Proof We must show that every object with the property $A_1 \cup A_2 \cup \cdots \cup A_k$ is counted once in the right side of (3). Note that objects which do not have the property $A_1 \cup A_2 \cup \cdots \cup A_k$ are not counted at all. Suppose an object has exactly s of the properties $A_1, A_2, \ldots, A_k$. For definiteness, suppose it has the properties $A_1, A_2, \ldots, A_s$ but not the properties $A_{s+1}, A_{s+2}, \ldots, A_k$. Then it will be counted in each of the s terms $N(A_i)$ for $i = 1, 2, \ldots, s$. It will be counted negatively in the terms $N(A_iA_j)$ if i, j is a 2-combination formed from the integers $1, 2, \ldots, s$. Then are $\binom{s}{2}$ such terms. It will be counted $\binom{s}{r}$ times in the term $(-1)^{r-1} N(A_{i_1} \cdots A_{i_r})$ as long as $r \leqslant s$ since there are $\binom{s}{r}$ r-combinations $(i_1, i_2, \ldots, i_s)$ which can be formed from the integers $1, \ldots, s$. Hence the total number of times the object is counted is

$$\binom{s}{1} - \binom{s}{2} + \binom{s}{3} - \cdots + (-1)^{s-1} \binom{s}{s}. \tag{4}$$

We must show that this expression is equal to 1. Expanding $(1 - 1)^s$ by the binomial theorem yields

$$0 = (1 - 1)^s = \binom{s}{0} - \binom{s}{1} + \binom{s}{2} - \binom{s}{3} + \cdots + (-1)^s.$$

Transposing $\binom{s}{0} = 1$ to the left side and multiplying by -1 we obtain an equation which states that (4) is equal to 1.

COROLLARY

$$N(A_1'A_2' \cdots A_k') = N - \sum N(A_i) + \sum N(A_iA_j) - \cdots + (-1)^r \sum N(A_{i_1} \cdots A_{i_r}) + \cdots + (-1)^k N(A_1 \cdots A_k),$$

where the summation is over all r-combinations formed from the integers $1, \ldots, k$.

The proof of this result follows immediately from the theorem and the corollary of theorem 1. The reader should try to establish this result directly by using the principle of inclusion and exclusion.

Using this more general result, example 4 can be generalized to yield some important theorems of number theory (see exercises 1, 2 and 3).

Example 5 Show that the number of partitions of n in which no integer occurs more than $r - 1$ times as a part is equal to the number of partitions of n into parts not divisible by r.

Before proving this result we illustrate its meaning by the examples on page 106.

Let a partition of n have the property A (B) if and only if at least one part is divisible by r (some part occurs at least r times). We must prove $N(A') = N(B')$. However it suffices to show that $N(A) = N(B)$ since this implies the required result by virtue of the fact that $N = N(A) + N(A') = N(B) + N(B')$. We will show that $N(A) = N(B)$ by applying theorem 3.

Let a partition of n have the property A_i if and only if the integer ir occurs as one of the parts. Of course there are only a finite number of A_i since $i \leqslant \left[\frac{n}{r}\right] = k$. It is not difficult to verify that $A = A_1 \cup A_2 \cup \cdots \cup A_k$ and so by the above theorem,

$$N(A) = N(A_1) + \cdots + N(A_k) - N(A_1A_2) - N(A_1A_3) \ldots \tag{5}$$

Similarly, let a partition of n have the property B_i if and only if the integer i occurs at least r times among the parts. Then $B = B_1 \cup \cdots \cup B_k$ and

$$N(B) = N(B_1) + \cdots + N(B_k) - N(B_1B_2) - N(B_1B_3) \ldots \tag{6}$$

$n = 10$, $r = 2$	Partitions									
each integer occurs once	10	9 + 1	8 + 2	7 + 3	6 + 4	7 + 1 + 2	6 + 1 + 3	5 + 2 + 3	1 + 2 + 3 + 4	5 + 4 + 1
parts not divisible by 2	9 + 1	7 + 3	5 + 5	7 + 1 + 1 + 1	5 + 1 + 1 + ⋯ + 1	5 + 3 + 1 + 1	1 + 1 + ⋯ + 1	3 + 3 + 3 + 1	3 + 3 + 1 + ⋯ + 1	3 + 1 + ⋯ + 1

$n = 6$, $r = 3$	Partitions						
each integer occurs $\leqslant 2$ times	6	5 + 1	4 + 1 + 1	4 + 2	3 + 3	3 + 2 + 1	2 + 2 + 1 + 1
parts not divisible by 3	5 + 1	4 + 2	2 + 2 + 2	4 + 1 + 1	2 + 2 + 1 + 1	2 + 1 + 1 + 1 + 1	1 + 1 + ⋯ + 1

We will now show that $N(A_i) = N(B_i)$. Given any partition

$$n = ir + x_1 + x_2 + \cdots + x_m$$

with the property A_i we can uniquely associate to it the partition

$$n = \overbrace{i + i + \cdots + i}^{r \text{ times}} + x_1 + x_2 + \cdots + x_m$$

with the property B_i, and conversely. Hence $N(A_i) = N(B_i)$ for all i. By the same reasoning, $N(A_iA_j) = N(B_iB_j)$, $N(A_iA_jA_k) = N(B_iB_jB_k)$ etc. Therefore all terms on the right side of (5) are equal to the corresponding terms of (6) and so $N(A) = N(B)$.

Example 6 (problème des ménages) In how many ways can n couples ($n \geqslant 3$)* be seated at a circular table so that men and women are in alternate places, but no husband sits on either side of his wife?

First seat the women at alternate places, designating them by the numbers $1, 2, \ldots, n$ in circular order. Denote the place between the ith and $(i + 1)$st ladies by the number i for $i = 1, 2, \ldots, n - 1$ and give the number n to the place between the nth lady and the first. Then the first husband can sit anywhere except in the nth or the first place and the ith husband anywhere except the $(i - 1)$st or the ith place for $i = 2, 3, \ldots, n$. If husband h_i sits in place i, then $h_1, h_2, \ldots, h_n$ is a permutation of $1, 2, \ldots, n$. Because of the restriction stated above, $h_i \neq i$ or $i + 1$ for $i = 1, 2, \ldots, n - 1$ and $h_n \neq n$ or 1. The number of such permutations, which is the solution to the problem, can be found by using the corollary to theorem 3. Let A_i be the property that $h_i = i$ or $i + 1$ for $i = 1, 2, \ldots, n - 1$ and A_n be the property that $h_n = n$ or 1. We must then calculate $\sum N(A_{i_1}A_{i_2} \cdots A_{i_r})$, the total number of ways that r of the properties A_i hold. Consider the sequence of symbols:

$$1_{h_n}1_{h_1}2_{h_1}2_{h_2}3_{h_2}3_{h_3}4_{h_3}4_{h_4} \cdots (n-1)_{h_{n-2}}(n-1)_{h_{n-1}}\, n_{h_{n-1}}n_{h_n}$$

arranged in a circle so that 1_{h_n} and n_{h_n} are consecutive. Each r-combination having no two consecutive symbols corresponds uniquely to an allowable value of r of the h_i's specified by the r properties A_i which hold and conversely. For example, $1_{h_n}\, 2_{h_1}\, 3_{h_2}\, 4_{h_3} \cdots r_{h_{r-1}}$ signifies that $h_n = 1$, $h_2 = 2, \ldots,$ $h_{r-1} = r$; while $1_{h_1}\, 2_{h_2}\, 3_{h_3} \cdots r_{h_r}$ signifies that $h_1 = 1$, $h_2 = 2, \ldots, h_r = r$. Two consecutive symbols cannot be picked as this would signify either that

* There is no solution if $n < 3$.

two husbands are sitting in the same chair or that one husband is sitting in two chairs. By the corollary to theorem 1.4.3, the number of such r-combinations is $\frac{2n}{2n-r}\binom{2n-r}{r}$. The number of permutations with the property $A_{i_1}A_{i_2}\cdots A_{i_r}$ with $h_{i_1}, h_{i_2}, \ldots, h_{i_r}$ fixed is $(n-r)!$ since the remaining h_i's can be arbitrary. Hence

$$\sum N(A_{i_1}A_{i_2}\cdots A_{i_r}) = \frac{2n}{2n-r}\binom{2n-r}{r}.$$

Therefore by the corollary to theorem 3, the solution of the "ménage" problem, M_n, is given by

$$M_n = n! - 2n(n-1)! + \cdots + (-1)^r \frac{2n}{2n-r}\binom{2n-r}{r}(n-r)! + \cdots + (-1)^n 2.$$

It can be shown that the M_n satisfy the recursion:

$$(n-2)\,M_n = n(n-2)\,M_{n-1} + nM_{n-2} + 4(-1)^{n+1}.$$

The formulas in theorems 1, 2 and 3 and their corollaries need not be memorized. They can easily be derived symbolically by replacing every property A' by $1 - A$ and multiplying out the resulting expression using the ordinary rules of multiplication. Note that $AB = BA, ABC = ACB = \cdots = CAB = CBA$ and $AA = AAA = AAAA = A$. The resulting expression is simplified by using the following conventions: $N(A + B) = N(A) + N(B)$, $N(-A) = -N(A)$, and $N(1) = N$. The integer 1 is used to denote a property which is possessed by all the objects under consideration. Note that $A \cup A' = 1$ and $N(A) + N(A') = N$. For example, the corollary to theorem 2 may be derived symbolically as follows;

$$\begin{aligned} N(A'B') &= N((1-A)(1-B)) = N(1 - A - B + AB) \\ &= N + N(-A) + N(-B) + N(AB) = N - N(A) - N(B) + N(AB). \end{aligned}$$

Example 7

$$\begin{aligned} N(AB'C'D) &= N(A(1-B)(1-C)D) \\ &= N(AD - ABD - ACD + ABCD) \\ &= N(AD) + N(-ABD) + N(-ACD) + N(ABCD) \\ &= N(AD) - N(ABD) - N(ACD) + N(ABCD). \end{aligned}$$

The above example can be generalized. If the properties $A_1, A_2, \ldots, A_n$ are relabeled as $B_1B_2, \ldots, B_k, C_1, C_2, \ldots, C_{n-k}$, then

$$N(B_1B_2 \cdots B_kC_1'C_2' \cdots C_{n-k}') = N(B_1B_2 \cdots B_k(1 - C_1)(1 - C_2) \cdots$$

$$(1 - C_{n-k})) = N(B_1B_2 \cdots B_k) - N(B_1B_2 \cdots B_kC_1) - \cdots$$

$$- N(B_1 \cdots B_kC_{n-k}) + N(B_nB_2 \cdots B_kC_1C_2) + \cdots$$

$$+ (-1)^{n-k} N(B_1 \cdots B\ C_1 \cdots C_{k-1}). \tag{7}$$

It can be shown that (7) is valid by using the principle of inclusion and exclusion or induction.

Quantitative statistical surveys are frequently conducted in which objects are classified according to their lack of or possession of certain attributes.

Example 8 Let the objects under consideration be a group of people. Suppose A represents the property of having blue eyes, B the property of being blond, and C the property of being male. Then the group of people can be separated into 8 classes corresponding to the properties:

$$ABC, A'BC, AB'C, ABC', A'B'C, A'BC', AB'C', A'B'C'. \tag{8}$$

For instance, a person is in the class corresponding to $A'B'C'$ if and only if the person is not blonde, not blue-eyed, and female.

The properties listed in (8) will be called *ultimate properties* and the corresponding classes *ultimate classes*. In general, if n properties A_i are given there are 2^n ultimate properties since each of the n properties in $B_1B_2 \cdots B_n$ can be filled by either A_i or A_i'. Every property can be expressed in terms of the ultimate properties. For example $A = AB \cup AB'$ since a person is blue-eyed if and only if the person is blue-eyed and blonde or blue-eyed and not blonde. Similarly

$$AB = ABC' \cup ABC, \qquad AB' = AB'C \cup AB'C';$$

hence

$$A = ABC' \cup ABC \cup AB'C \cup AB'C'.$$

The general case can be treated in a similar manner.

Next observe that each object must belong to exactly one of the ultimate classes. Since two ultimate classes have no common objects,

$$N(AB) = N(ABC') + N(ABC),$$
$$N(A) = N(ABC') + N(ABC) + N(AB'C) + N(AB'C')$$

etc. Hence the number of objects with any property can be expressed as the sum of numbers of objects with ultimate properties.

Every class of objects always contains a nonnegative number of objects; in particular, the ultimate classes contain a nonnegative number of objects. This criterion is used as a test for the consistency of data in statistical surveys. The number of elements in the ultimate classes is calculated by means of (7). Note that even if the number of objects in all ultimate classes are positive the data need not be consistent. In other words, the above criterion is necessary but not sufficient for consistency.

Example 9 A cigarette company surveyed $N = 100{,}000$ people and claimed to have obtained the following results:

number of males = 50,000, number of smokers = 80,000, number with cancer = 1,000, number of male smokers = 45,000, number of males with cancer = 700, number of smokers with cancer = 450, number of male smokers with cancer = 149.

Are these figures consistent?

Let M be the property of being a male, C the property of having cancer, and S the property of being a smoker. Then

$$\begin{aligned} N(M'S'C) &= N((1 - M)(1 - S)C) \\ &= N(C) - N(MC) - N(SC) + N(MSC) \\ &= 1000 - 700 - 450 + 149 = -1 \end{aligned}$$

and so the data is inconsistent.

Exercises

1 Let $p_1, \ldots, p_m$ be positive integers, no two of which have a common factor greater than 1. Show that if N is a positive integer, then the number of positive integers less than N which are not divisible by any of the numbers $p_1, p_2, \ldots, p_m$ is

$$N - \sum \left[\frac{N}{p_i}\right] + \sum \left[\frac{N}{p_i p_j}\right] - \sum \left[\frac{N}{p_i p_j p_k}\right] + \cdots + (-1)^m \left[\frac{N}{p_1 p_2 \cdots p_m}\right].$$

2 How many positive integers $\leqslant 1260$ are relatively prime to 1260?

3 Let $N = p_1^{i_1} p_2^{i_2} \cdots p_m^{i_m}$ be the prime factorization of the positive integer $N \neq 1$. Then

$\phi(N)$, the number of positive integers less than N and relatively prime* to N is given by

$$\phi(N) = N - \frac{N}{p_1} - \cdots - \frac{N}{p_m} + \frac{N}{p_1p_2} + \cdots + \frac{N}{p_{m-1}p_m}$$
$$- \frac{N}{p_1p_2p_3} - \cdots (-1)^m \frac{N}{p_1p_2\cdots p_m}$$
$$= N\left(1 - \frac{1}{p_1}\right)\left(1 - \frac{1}{p_2}\right)\cdots\left(1 - \frac{1}{p_m}\right).$$
$$\phi(1) = 1$$

4 Three newspapers A, B, and C are published in a certain city. The following results were obtained in a survey of the adult population: 20% read A, 16% read B, 14% read C, 8% read both A and B, 5% read both A and C, 4% read B and C, 2% read all three. What percentage reads at least one of the papers? Of those who read at least one of the papers, what percentage reads both A and B?

5 The following data is a census of England and Wales in 1891. Total population = 29,002,252. Number of blind = 23,467, number of deaf mutes = 14,192, number of mentally diseased = 97,383, number of blind deaf mutes = 107, number of mentally diseased deaf mutes = 525, number of blind mentally diseased = 405, number of blind, mentally diseased deaf mutes = 25. Find the number of people in each ultimate class.

6 How many integers from 1 to 10^6 inclusive are neither perfect squares, perfect cubes, nor perfect fourth powers?

7 How many solutions are there of the equation

$$x_1 + x_2 + x_3 + x_4 = 30$$

in positive integers with $x_1 \leqslant 5$, $x_2 \leqslant 7$, $x_3 \leqslant 16$, and $x_4 \leqslant 19$? (Hint: Let the "objects" be the solutions of the equation in positive integers—for example, $x_1 = x_2 = x_3 = 5$, $x_4 = 15$ is an object. A solution has the property A in case $x_1 > 5$, property B in case $x_2 > 7$ etc. Find $N(A'B'C'D')$.

8 In a test of some circuits of rocketships, 10 were found to be defective; seven had worn-out transistors, five had faulty connections, and four had broken wires. One had both a worn-out transistor and a faulty connection but good wires; one had a faulty connection and a broken wire but good transistors; two had both worn-out transistors and broken wires but good connections; and three had only worn-out transistors.

a) How many had all three defects?
b) How many had broken wires only?

2 Counting objects with *m* properties

Imagine that hats in a checkroom are mixed and distributed at random to the patrons. We might be interested in the number of ways in which exactly 1 person gets his own hat. For simplicity, assume that there are only 4 people.

* Two positive integers are relatively prime if their greatest common divisor is 1.

Let A_i be the property that patron i gets his own hat and let the objects be the various possible ways of distributing the hats. Then the number of ways is given by

$$\begin{aligned} &N(A_1A_2'A_3'A_4') + N(A_1'A_2A_3'A_4') + N(A_1'A_2'A_3A_4') + N(A_1'A_2'A_3'A_4') \\ &\quad = N(A_1(1 - A_2)(1 - A_3)(1 - A_4)) + \cdots \\ &\qquad + N((1 - A_1)(1 - A_2)(1 - A_3)A_4) \\ &\quad = N(A_1) + N(A_2) + N(A_3) + N(A_4) - 2[N(A_1A_2) \\ &\qquad + (A_1A_3) + N(A_1A_4) + N(A_2A_3) + N(A_2A_4) \\ &\qquad + N(A_3A_4)] + 3[N(A_1A_2A_3) + N(A_1A_2A_4) \\ &\qquad + N(A_1A_3A_4) + N(A_2A_3A_4)] - 4N(A_1A_2A_3A_4). \end{aligned} \tag{1}$$

To save writing, we introduce the following abbreviations:

$$\begin{aligned} &S_0 = N,\ S_1 = \sum N(A_i),\ S_2 = \sum N(A_iA_j),\ \ldots,\ S_r = \sum N(A_{i_1} \cdots A_{i_r}),\ \ldots, \\ &\qquad S_n = N(A_1 \cdots A_n) \end{aligned} \tag{2}$$

where the summation in S_r is over all r-combinations $A_{i_1}A_{i_2} \cdots A_{i_r}$ formed from the n letters $A_1, A_2, \ldots, A_n$. (1) can be written as

$$S_1 - \binom{2}{1} S_2 + \binom{3}{1} S_3 - \binom{4}{1} S_4. \tag{1'}$$

$N(A_i) = 3!$ since person i gets his hat while the other hats can be arranged in $3!$ ways; hence $S_1 = 4(3!)$. Two people can get their own hats in $2!$ ways and hence $S_2 = \binom{4}{2} 2!$. Since $N(A_iA_jA_k) = 1$, $S_3 = \binom{4}{3}$; $S_4 = 1$. Substituting these values in (1′), the number of ways is

$$4! - 2\frac{4!}{2!\,2!}\,2! + 3\frac{4!}{3!} - 4 = 4!\left(1 - 1 + \frac{1}{2!} - \frac{1}{3!}\right) = 8.$$

The number of ways exactly 2 patrons get their own hats is

$$\begin{aligned} &N(A_1A_2A_3'A_4') + N(A_1A_2'A_3A_4') + N(A_1A_2'A_3'A_4) + N(A_1'A_2A_3A_4') \\ &\qquad + N(A_1'A_2A_3'A_4) + N(A_1'A_2'A_3A_4) \\ &\quad = N(A_1A_2(1 - A_3)(1 - A_4)) + \cdots \\ &\qquad + N((1 - A_1)(1 - A_2)A_3A_4) \\ &\quad = S_2 - \binom{3}{2} S_3 + \binom{4}{2} S_4. \end{aligned} \tag{3}$$

By exactly the same method, the number of ways exactly 3 patrons get their own hat is

$$S_3 - \binom{4}{3} S_4. \tag{4}$$

To gain further insight into the general result we find the number of objects which have exactly 1 of the five properties $A_1, A_2, \ldots, A_5$, namely,

$$\begin{aligned} N(A_1A_2'A_3'A_4'A_5') &+ N(A_1'A_2A_3'A_4'A_5') + \cdots N(A_1'A_2'A_3'A_4'A_5) \\ &= N(A_1(1 - A_2)(1 - A_3)(1 - A_4)(1 - A_5)) + \cdots \\ &\qquad + N((1 - A_1)(1 - A_2)(1 - A_3)(1 - A_4)A_5) \\ &= S_1 - \binom{2}{1} S_2 + \binom{3}{1} S_3 - \binom{4}{1} S_4 + \binom{5}{1} S_5. \end{aligned} \tag{5}$$

(1′), (3), (4) and (5) suggest the following result.

THEOREM 2 *For any integer* m ($1 \leqslant m \leqslant n$) the number of objects $N_{[m]}$ that have exactly m of the n properties is given by*

$$\begin{aligned} N_{[m]} = S_m &- \binom{m+1}{m} S_{m+1} + \binom{m+2}{m} S_{m+2} - \cdots \\ &+ (-1)^{n-m} \binom{n}{m} S_n. \end{aligned} \tag{6}$$

Proof Suppose that an object has r of the given properties. If $r < m$ this object is not counted in (6). If $r = m$ the object is counted only once, namely, by $N(A_{i_1} \cdots A_{i_m})$ where $A_{i_1} \cdots A_{i_m}$ are the m properties possessed by the object. Finally if $r > m$ the object is counted $\binom{r}{m}$ times in S_m, $\binom{r}{m+1}$ times in S_{m+1} and so on until S_r is reached. The object is not counted in S_{r+1}, $S_{r+2}, \ldots, S_n$ since it has exactly r properties. Hence the object contributes

$$\begin{aligned} \binom{r}{m} &- \binom{m+1}{m}\binom{r}{m+1} + \binom{m+2}{m}\binom{r}{m+2} - \cdots \\ &+ (-1)^{r-m} \binom{r}{m}\binom{r}{r} \end{aligned} \tag{7}$$

* The theorem also holds for $m = 0$. According to the corollary to theorem 1.3, $N_{[0]} = N - S_1 + S_2 - \cdots + (-1)^n S_n$.

to (6). Using the identity

$$\binom{m+i}{m}\binom{r}{m+i} = \binom{r}{m}\binom{r-m}{i}, \tag{8}$$

(7) can be replaced by

$$\binom{r}{m}\left[1 - \binom{r-m}{1} + \binom{r-m}{2} - \cdots + (-1)^{r-m}\right].$$

However the expression in brackets is just the expansion of $(1 - 1)^{r-m}$ and so is zero. Hence only objects with exactly m properties are counted by (6).

*Example 1** Let a deck of n distinct cards be laid out in a row and another similar well-shuffled deck be dealt out underneath the first. Then in a random matching of the two equivalent decks the number of ways of having exactly m matches is given by

$$N_{[m]} = \frac{n!}{m!}\left(1 - 1 + \frac{1}{2!} - \frac{1}{3!} + \cdots + (-1)^{n-m}\frac{1}{(n-m)!}\right).$$

We apply formula (6). Let A_i be the property that the cards in the ith position match and let the objects under consideration be all possible arrangement of the cards. $N(A_{i_1} \cdots A_{i_m}) = (n - m)!$ since there are m matches while the remaining $n - m$ cards can be in any order. This result is independent of which cards match and so $S_m = \binom{n}{m}(n - m)!$, since there are $\binom{n}{m}$ terms in S_m corresponding to the number of m-combinations $A_{i_1} \cdots A_{i_m}$ which can be formed from the n letters $A_1 \cdots A_n$. By exactly the same reasoning, $N(A_{i_1} \cdots A_{i_r}) = (n - r)!$ and $S_r = \binom{n}{r}(n - r)!$. Substituting these values in (6) yields

$$\begin{aligned} N_{[m]} = \binom{n}{m}(n-m)! - \binom{m+1}{m}\binom{n}{m+1}(n-m-1)! \\ + \binom{m+2}{m}\binom{n}{m+2}(n-m-2)! + \cdots \\ + (-1)^{n-m}\binom{n}{m}. \end{aligned}$$

* This problem is known as the "problème des rencontres". If $m = 0$ the permutation is called a *derangement*.

By (8), the above expression reduces to

$$N_{[m]} = \binom{n}{m}\left[(n-m)! - \binom{n-m}{1}(n-m-1)! + \binom{n-m}{2} \times \right.$$

$$\left. \times\, (n-m-2)! - \binom{n-m}{3}(n-3)! + \cdots + (-1)^{n-m}\right]$$

$$= \frac{n!}{m!}\left[1 - 1 + \frac{1}{2!} - \frac{1}{3!} + \cdots + \frac{(-1)^{n-m}}{(n-m)!}\right].$$

Example 2 In how many ways can r different balls be placed in n different unordered cells so that exactly m cells are empty? (cf. theorem 3.3.5.)

Let A_i be the property that cell i is empty and let the objects under consideration be all possible distributions of the balls in n cells. Then $N(A_{i_1}, \ldots, A_{i_k}) = (n-k)^r$ since there are $n-k$ cells in which r balls must be distributed. Since this number is independent of which cells are empty, $S_k = \binom{n}{k}(n-k)^r$ as there are $\binom{n}{k}$ terms of the type $N(A_{i_1} \cdots A_{i_k})$ in S_k. Hence from (6),

$$N_{[m]} = \binom{n}{m}\sum_{i=0}^{n-m}(-1)^i\binom{n-m}{i}(n-m-i)^r.$$

A formula for the number of objects with at least m of n given properties can easily be obtained from theorem 1.

THEOREM 2 *The number of objects N_m which possess m or more of the properties $A_1, A_2, \ldots, A_n$ is given by*

$$N_m = S_m - \binom{m}{m-1}S_{m+1} + \binom{m+1}{m-1}S_{m+2} - \cdots + (-1)^{n-m}\binom{n-1}{m-1}S_n. \tag{9}$$

Proof An object has at least m properties if and only if it has exactly m or exactly $m+1$ or $\cdots$ or exactly n of the properties. Therefore

$$N_m = N_{[m]} + N_{[m+1]} + \cdots + N_{[n]}.$$

From (6) we see that the coefficient of S_{m+r} is

$$(-1)^r\binom{m+r}{m} + (-1)^{r-1}\binom{m+r}{m+1} + (-1)^{r-2}\binom{m+r}{m+2} + \cdots$$
$$+ (-1)^{r-r}\binom{m+r}{m+r}$$
$$= (-1)^r\left(\left[\binom{m+r-1}{m-1} + \binom{m+r-1}{m}\right] - \left[\binom{m+r-1}{m}\right.\right.$$
$$\left. + \binom{m+r-1}{m+1}\right]$$
$$\left. + \left[\binom{m+r-1}{m+1} + \binom{m+r-1}{m+2}\right] + \cdots\right)$$
$$= (-1)^r\binom{m+r-1}{m-1}.$$

Example 3 In how many ways can at least 3 heads be obtained upon tossing 5 coins?

Let A_i be the property that the ith coin falls heads and let the objects under consideration be the outcomes upon tossing 5 coins. Since $N(A_{i_1}A_{i_2}A_{i_3}) = 2^2$ (as each of the two remaining coins can be heads or tails), we have $S_3 = \binom{5}{3}2^2$. Similarly,

$$S_4 = \binom{5}{4}2 \quad \text{and} \quad S_5 = \binom{5}{5}; \quad \text{hence from (9),}$$

$$N_3 = \binom{5}{3}2^2 - \binom{5}{4}2\binom{3}{2} + \binom{5}{5}\binom{4}{2} = 40 - 30 + 6 = 16.$$

Exercises

1 Ten pairs of gloves are in a drawer. Four gloves are selected at random. Find the number of ways that at least one pair is among the gloves selected.

2 Find the number of ways that a coin falls tails at least 3 times in a succession upon tossing a coin 5 times.

3 A simple code is made by permuting the letters of the alphabet, with every letter replaced by a different one. How many codes can be made in this way?

4 The number of ways in which 8 rooks can be placed on a chessboard so that none can take another and that none stands on the white diagonal is

$$8!\left(1 - 1 + \frac{1}{2!} - \frac{1}{3!} + \frac{1}{4!} - \cdots + \frac{1}{8!}\right).$$

5 Five dice are thrown. Find the number of ways that at least 3 of them show the same face.

6 Each box of cereal contains one picture card. Suppose there are 5 cards of each kind and n different kinds of pictures. In how many ways can $r \geqslant n$ boxes contain all the different kinds of pictures?

7 The number of terms containing one or more diagonal elements in the general expansion of a determinant of order n is

$$n!\left(1 - \frac{1}{2!} + \frac{1}{3!} - \cdots + (-1)^{n-1}\frac{1}{n!}\right).$$

8 Verify that the number of ways of placing r different balls in r different unordered cells so that exactly m cells are occupied by exactly k balls is

$$N_{[m]}(k) = \frac{(-1)^m n!\, r!}{m!} \sum_j (-1)^j \frac{(n-k)^{r-jk}}{(j-m)!\,(n-j)!\,(r-jk)!\,(k!)^j};$$

where the summation extends over those $j \geqslant m$ for which $j \leqslant n$ and $jk \leqslant r$.

9 Let $H(x)$ b the generating function of the sequence $\{N_{[m]}\}$ for $m = 1, 2, \ldots, n$. Show that

$$H(x) = \sum_{j=0}^{n} S_j (x-1)^j.$$

(Hint:
$$\begin{aligned}
H(x) &= [S_0 - S_1 + S_2 - \cdots + (-1)^n S_n] \\
&\quad + \left[S_1 - \binom{2}{1} S_2 + \binom{3}{2} S_3 - \cdots + (-1)^{n-1}\binom{n}{n-1} S_n\right] x \\
&\quad + \cdots \\
&\quad + S_n x^n \\
&= S_0 \\
&\quad + S_1 [x-1] \\
&\quad + S_z \left[x^2 - \binom{2}{1} x + 1\right] \\
&\quad + \cdots \\
&\quad + S_n \left[x^n - \binom{n}{1} x^{n-1} + \binom{n}{2} x^{n-2} + \cdots + (-1)^n\right] \\
&= \Sigma S_j (x-1)^j.
\end{aligned}$$

3. Rook polynomials

A *rook* is a piece in the game of chess which "captures" (or "takes") on both rows or columns of a chessboard. In other words, if an opponent's piece is in the same row or column as your rook and no other piece is between them it may be removed from the board. The problem of nontaking rooks is to enumerate the number of ways of placing n identical rooks on a chessboard so that no rook can be captured by any other rook. As shown below, the solution of this problem is equivalent to finding the number of permutations of objects when there are arbitrary restrictions on the positions they can occupy.

Example 1

1 On a regular 8×8 chessboard, there are 64 ways to place one (nontaking) rook. Two nontaking rooks can be placed in $\binom{8}{2}\,{}_8P_2$ positions since there are $\binom{8}{2}$ ways to choose two rows and then ${}_8P_2$ ways to choose two squares from the chosen rows. At most 8 nontaking rooks can be placed on the board and there are $8!$ ways to do so.

2 More generally, it will be necessary to place nontaking rooks on chessboards of arbitrary shapes and sizes as illustrated in figure 1.

Figure 1

On chessboard C_1 there are 3 ways to place one nontaking rook, 1 way to place two nontaking rooks and there is no way to place more than two nontaking rooks. On C_2 there are 5 ways to place one nontaking rook, 6 ways to place two nontaking rooks, 1 way to place three nontaking rooks and no way to place more than three nontaking rooks.

DEFINITION 1 *Let $r_k(C)$ denote the number of ways of placing k nontaking rooks on a chessboard C with n cells. The generating function*

$$R(x, C) = \sum_{k=0}^{n} r_k(C)\, x^k$$

of the sequence $\{r_k(C)\}$ is called the rook polynomial *of the given chessboard.*

When there is no danger of confusion, $r_k(C)$ will be denoted by r_k and the corresponding rook polynomial by $R(x)$. Notice that $R(x)$ is a polynomial of degree at most n, where n is the number of cells of the chessboard, because it is impossible to place more than n nontaking rooks on a chessboard of n cells.

Example 2

1 The rook polynomial for an ordinary 8×8 chessboard is

$$R(x) = \sum_{k=0}^{8} \binom{8}{k} \frac{8!}{(8-k)!} x^k.$$

2 The rook polynomials in part 2 of example 1 are:

$$R(x, C_1) = 1 + 3x + x^2;$$
$$R(x, C_2) = 1 + 5x + 6x^2 + x^3.$$

The rook polynomial for more complicated chessboards can sometimes be obtained by means of the following *expansion formula.*

THEOREM 1 *Suppose that on a given chessboard C a cell is selected. Let C_r denote the chessboard obtained from C by deleting the row and column containing the selected cell. Let C_s denote the chessboard obtained from C by deleting the selected cell. Then*

$$R(x, C) = xR(x, C_r) + R(x, C_s). \tag{1}$$

Proof The ways of placing k nontaking rooks on C can be divided into two classes; those that have a rook in the selected cell and those that do not have a rook in the selected cell. The number of ways in the first class is equal to $r_{k-1}(C_r)$ and the number of ways in the second class is equal to $r_k(C_s)$. Hence

$$r_k(C) = r_{k-1}(C_r) + r_k(C_s).$$

This last relation, written in terms of generating functions, is just (1).

The method of finding the rook polynomial for a chessboard, by repeated applications of the expansion formula, is illustrated in the following example.

Example 3

1 A pair of parentheses around a chessboard will be used to denote the rook polynomial of the board. Hence the expansion formula applied to C_1 in Fig. 1 (with respect to the cell in the lower left corner) can be written as

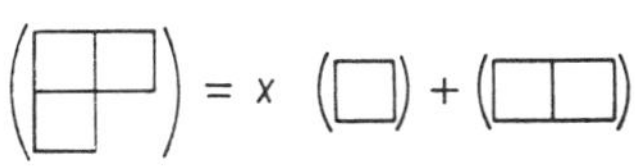 $= x(1 + x) + (1 + 2x) = 1 + 3x + x^2.$

2 The expansion formula when applied to C_2 in Fig. 1 yields

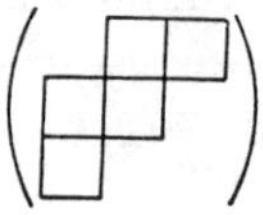 $= x$ 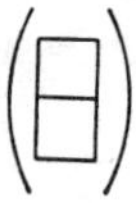 $+$ 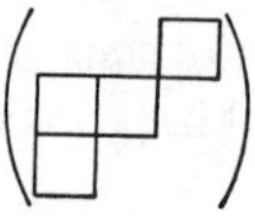

$= x\,(1 + 2x) + x$ 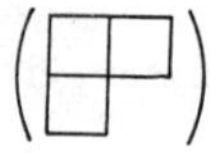 $+$ 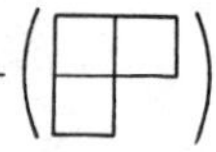

$$= x + 2x^2 + x\,(1 + 3x + x^2) + (1 + 3x + x^2)$$

$$= 1 + 5x + 6x^2 + x^3.$$

Two subboards are *disjunct* if no cell of one subboard is in the same row or in the same column as any cell of the other subboard. It is frequently necessary to determine the rook polynomial of a chessboard consisting of two disjunct subboards.

Example 4

1 In part 2 of example 3, the chessboard

$C =$ 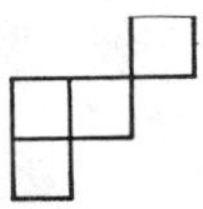

consists of the two disjunct subboards

$C_1 =$ 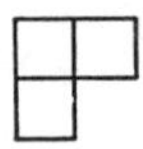 and $C_2 =$ .

2 The chessboard

$C =$ 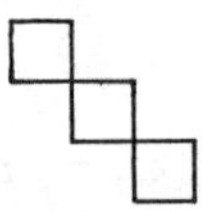

consists of 3 disjunct chessboards $C_1 = C_2 = C_3 =$ .

The following theorem greatly simplifies the calculation of the rook polynomial for such chessboards.

THEOREM 2 *If a chessboard C consists of two disjunct subboards C_1 and C_2 then*

$$R\,(x, C) = R\,(x, C_1)\,R\,(x, C_2). \tag{2}$$

Proof Noting that the way rooks are placed on C_1 is completely independent of the way rooks are placed on C_2 it follows that

$$r_k(C) = \sum_{j=0}^{k} r_j(C_1)\, r_{k-j}(C_2) \tag{3}$$

since k rooks can be placed on C by placing j rooks on C_1 and then $k - j$ on C_2 ($j = 0, 1, 2, \ldots, k$). Using the definition of the product of two generating functions (see page 51), the result (2) follows from (3).

As an application of theorem 2, the rook polynomial for the chessboard in part 1 of example 4 is

$$R(x, C) = R(x, C_1)\, R(x, C_2) = (1 + 3x + x^2)(1 + x)$$

and the rook polynomial for the chessboard in part 2 is

$$R(x, C) = R(x, C_1)\, R(x, C_2)\, R(x, C_3) = 1 + x^3 .$$

The distribution of k out of m distinct objects into n distinct positions ($k \leqslant n$) is equivalent to placing k nontaking rooks on an $m \times n$ chessboard. For example, suppose a family consists of $m = 3$ adults (objects) a, b, and c. They are considering the purchase of $n = 4$ cars (positions) denoted by 1, 2, 3, and 4. There is only enough money for two of them to buy one car each ($k = 2$). (Of course, not the same car.) A distribution (way of purchasing cars) can be represented in the form of a chessboard as illustrated in Fig. 2. The rows correspond to the people and the columns correspond to the cars.

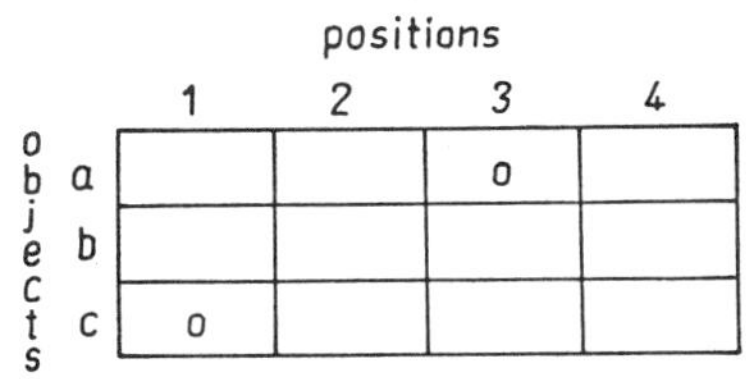

Figure 2

A circle in cell (x, j) indicates that the person x buys car j. Thus Fig. 2 indicates that a buys car 3 and c buys car 1. Since a person cannot buy more than one car and two people cannot buy the same car, there will never be more than one circle in a row or column.

If there are forbidden positions for the objects, X's are placed in the cells corresponding to the forbidden positions for each object. Circles can only be placed in unmarked cells; this is equivalent to placing nontaking rooks on the resulting unmarked chessboard. This procedure is illustrated in the following example.

Example 5

1 The number of derangements* of the integers 1, 2, 3, 4 is just r_4 for the unmarked chessboard in Fig. 3 (i.e.) the chessboard obtained when all the cells marked with an X are omitted),

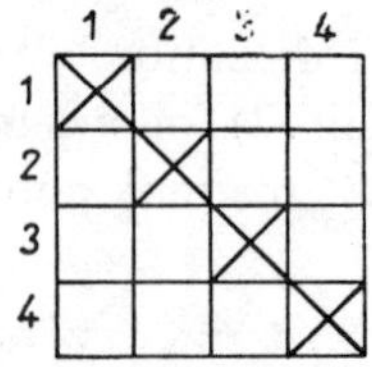

Figure 3

2 A computer dating service wants to arrange dates for five men 1, 2, 3, 4 and 5 and four women a, b, c and d. After analyzing psychological data, the computer decides that 1 should not be matched with c, 2 should not be matched with a and b, 3 and 4 should not be matched with d and 5 should not be matched with c. Find the number of ways to arrange four dates.

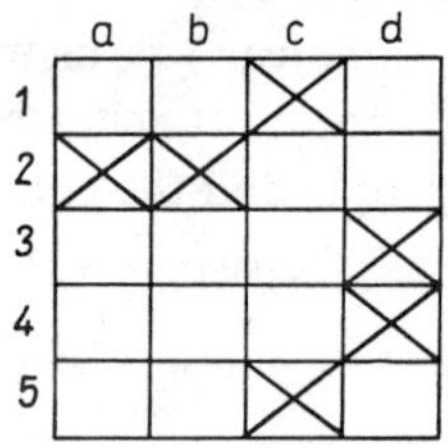

Figure 4

This a problem of permutations with restricted positions. That is, it is a 4-permutation of 5 numbers such that 1 is not in the third position, 2 is not in the first or second position, etc. The forbidden positions are marked with X's in Fig. 4. The number of ways of arranging four dates is just r_4 for the unmarked chessboard.

Frequently, the rook polynomial for the chessboard consisting of only those cells marked with an X (the chessboard of forbidden positions) is easier to find than the rook polynomial for the unmarked chessboard. For instance, in part 1 of example 5, the rook polynomial for the chessboard of

* No integer appears in its natural position.

forbidden positions is

$$R_1(x) = (1 + x)^4 = 1 + 4x + \binom{4}{2} x^2 + \binom{4}{3} x^3 + x^4 \tag{4}$$

and in part 2, the rook polynomial for the chessboard of forbidden positions is

$$R_2(x) = (1 + 2x)^3 = 1 + 6x + 12x^2 + 8x^3 \tag{5}$$

By using the principle of inclusion and exclusion, these rook polynomials can be used to find the number of permutations in which no object is in a forbidden position.

THEOREM 3 *The number of permutations of n objects in which no object is in a forbidden position is*

$$\sum_{k=0}^{n} (-1)^k r_k (n - k)!$$

where r_k is the number of ways of placing k nontaking rooks on the chessboard of forbidden positions.

Proof Let a permutation have the property A_k if the kth object is in a forbidden position. Then the number of permutations in which no object is in a forbidden position is $N(A_1'A_2' \cdots A_n')$. To apply the corollary of theorem 1.3 we must evaluate $\Sigma N(A_{i_1} \cdots A_{i_k})$, the total number of permutations with at least k objects in forbidden positions. This sum is equal to $r_k (n - k)!$ since k of the n objects can be placed in forbidden positions in r_k ways and the remaining $n - k$ objects can be placed in $n - k$ positions in $(n - k)!$ ways. The theorem follows immediately from this result upon applying the corollary to theorem 1.3.

Using theorem 3, the numbeı of derangements in part 1 of example 5 is

$$4! - \frac{4!}{3!\,1!}(3!) + \frac{4!}{2!\,2!}(2!) - \frac{4!}{3!\,1!}(1!) + 1$$

$$= 4!\left(1 - \frac{1}{1!} + \frac{1}{2!} + \frac{1}{3!} + \frac{1}{4!}\right)$$

since the r_k are the coefficients of x^k in (4) (cf. example 2.1).

To solve the problem in part 2 of example 5 by means of theorem 3, add another column of 5 cells to the chessboard on Fig. 4 and mark this column n. A circle in cell i ($i = 1, 2, 3, 4, 5$) of column n denotes the fact that the ith boy has no date. Hence the problem has been reduced to finding the number permutations of 5 objects in which no object is in a forbidden position. Using the r_k in (5), theorem 3 yields

$$5! - 6(4!) = 12(3!) - 8(2!) = 32$$

ways of arranging four dates.

The following generalization of theorem 3 is obtained by applying theorem 2.1.

THEOREM 4 *The number of permutations of n objects in which exactly m of the objects are in forbidden positions is*

$$N_{[m]} = r_m (n - m)! - \binom{m+1}{1} r_{m+1} (n - m - 1)! + \cdots + (-1)^{n-m} \binom{n}{n-m} r_n,$$

where r_k is the number of ways of placing k nontaking rooks on the chessboard of forbidden positions.

The generating function for the sequence $\{N_{[m]}\}$ for $m = 1, 2, \ldots, n$ can easily be obtained from exercise 2.9. It is

$$H(x) = \sum_{j=0}^{n} S_j (x-1)^j = \sum_{j=0}^{n} r_j (n-j)! (x-1)^j.$$

Since an object in a forbidden position is said to be a *hit*, $H(x)$ is called the *hit polynomial.* In part 2 of example 5, the hit polynomial is

$$\begin{aligned} H(x) &= 5!(x-1)^0 + 6(4!)(x-1)^1 + 12(3!)(x-1)^2 + 8(2!)(x-1)^3 \\ &= 32 + 48x + 24x^2 + 16x^3. \end{aligned}$$

Hence there are 32 ways to arrange dates with no incompatible couple, 48 ways with exactly one incompatible couple, 24 ways with exactly two incompatible couples and 16 ways with exactly 3 incompatible couples. There is no way to arrange four dates so that all the couples are incompatible.

Exercises

1 In how many ways can four rooms be painted four different colors: pink, white, yellow and green, if the first room cannot be painted green, the second cannot be painted white or yellow, the third cannot be painted pink or yellow and the fourth cannot be painted white?

2 Find the number of permutations of the letters a, a, b, b, c, c so that no a appears in the first and third position; no b appears in the second position and no c appears in the fifth position.

3 A pair of distinct dice are rolled six times. Suppose that the combinations (2,1), (3,2), (2,3), (4,4), (4,5) and (6,6) did not occur. Find the number of ways that all faces of the two dice could have appeared.

4 Find the rook polynomial for the following chessboard:

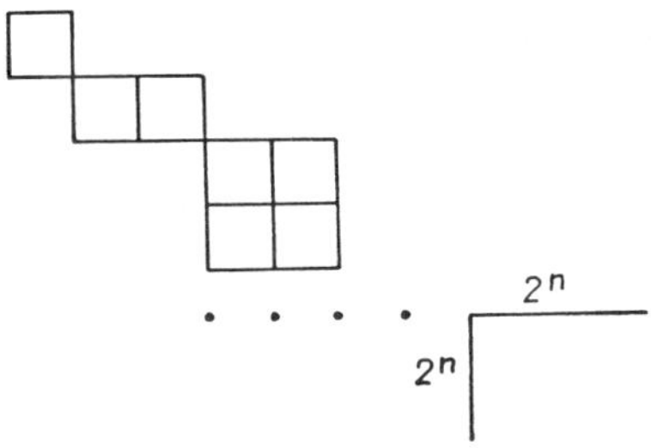

5 Two consecutive integers i, j in a permutation $(\ldots, i, j, \ldots)$ of the first n positive integers form an *ascent* if $j > i$. For example 1, 2, 3, 4, 5 has four ascents, 3, 1, 5, 4, 2 has one ascent and 5, 4, 3, 2, 1 has no ascents. Let $n = 5$ and consider a 5×5 chessboard whose columns and rows are labelled 1, 2, 3, 4, 5. Interpret a circle in cell (i, j) to mean that i is immediately to the left of j in the permutation. Find the chessboard consisting of forbidden cells if a permutation is to have no ascents. Find the corresponding hit polynomial and show that the coefficient of x^k is the number of permutations with k ascents for $k = 0, 1, 2, 3, 4$.

CHAPTER V

Application of combinatorial analysis to probability theory

IN THIS CHAPTER we illustrate some applications of combinatorial analysis to probability theory. We only consider situations in which there are a finite number of equally likely outcomes so that equation 1.1.1 is applicable.

1 Suppose an urn contains 10 white balls and 5 black balls. What is the probability of drawing 2 white balls and 1 black ball if 3 balls are selected at random?

First it is intuitively obvious that the chance of drawing the 3 balls is unchanged if we number the black and white balls. This allows us to distinguish two balls which appeaı similar. Second we should obtain the same result whether we choose the balls (a) one by one (without replacement) or (b) draw 3 balls at a time.

(a) Let W_i $(i = 1, \ldots, 10)$ denote a white ball and B_j $(j = 1, \ldots, 5)$ a black ball. We now have 15 different balls. There are 15 choices for the first ball; once this ball has been selected 14 choices for the second ball; similarly 13 choices for the third ball. Hence there are $15 \cdot 14 \cdot 13$ possible outcomes. 2 white balls and 1 black ball can be drawn in 3 ways—namely, BWW, WBW, WWB corresponding to the fact that the black ball can be the first or second or third ball chosen. Each of these ways yields $10 \cdot 9 \cdot 5$ favorable cases as there are 10 choices for the first white; once this has been selected 9 choices for the second white and finally 5 choices for the black. Hence the number of favorable cases is $3\,(10 \cdot 9 \cdot 5)$. The required probability is:

$$\frac{3\,(10 \cdot 9 \cdot 5)}{15 \cdot 14 \cdot 13} = \frac{45}{91}.$$

(b) If we draw 3 balls at a time, the order in which the balls are drawn does not count. Hence the total number of cases is $\binom{15}{3}$. There are $\binom{10}{2}$ choices for the two white balls and $\binom{5}{1}$ choices for the black ball; accordingly, there are $\binom{10}{2}\binom{5}{1}$ favorable cases. Hence the required probability is

$$\binom{10}{2}\binom{5}{1}\Big/\binom{15}{3} = \frac{45}{91}$$

which is the same as (a).

2 a) Suppose two dice are thrown. Find the probabilities of the various totals which can be obtained.

b) n different dice are thrown*. Find the probability that the n dice total r ($n \leqslant r \leqslant 6n$).

(a) Again we assume that the two dice are different; one is red and the other is green. By writing down the various alternatives it is not difficult to construct the following table.

Total	2	3	4	5	6	7	8	9	10	11	12
No. of ways	1	2	3	4	5	6	5	4	3	2	1
Probability	$\frac{1}{36}$	$\frac{2}{36}$	$\frac{3}{36}$	$\frac{4}{36}$	$\frac{5}{36}$	$\frac{6}{36}$	$\frac{5}{36}$	$\frac{4}{36}$	$\frac{3}{36}$	$\frac{2}{36}$	$\frac{1}{36}$

For example, a total of 6 can be obtained as follows: (1, 5), (5, 1), (2, 4), (4, 2), (3, 3) where the first number of each pair is the number on the red die. The total number of ways can be obtained by adding together all the numbers in the second row of the above table. Alternatively, there are 6 numbers on each die and so there are $6 \times 6 = 36$ possible outcomes.

(b) For large values of r and n we cannot list all the alternatives. We use the method of generating functions. Consider

$$(x^1 + x^2 + \cdots + x^6)^n. \tag{1}$$

The coefficient of x^r is the number of ways of totalling r since it is the number of terms of the form $x^{i_1}x^{i_2}\cdots x^{i_n}$ where $i_1 + i_2 + \cdots + i_n = r$ and each $i = 1$ or 2 or $\cdots$ or 6. The coefficient of x^r in (1) is the same as the coefficient of x^r in $x^n(1-x)^n(1-x^6)^n$ for $n \leqslant r \leqslant 6n$; this is

$$\sum_{k=0}^{\left[\frac{r-n}{6}\right]} (-1)^k \binom{n}{k}\binom{r-6k-1}{r-6k-n}.$$

* Note that this is the same as throwing 1 die n times.

The total number of ways is the sum of all the coefficients; this can easily be obtained by setting $x = 1$ in (1). Hence the probability

$$6^{-n} \sum_{k=0}^{\left[\frac{r-n}{6}\right]} (-1)^k \binom{n}{k} \binom{r-6k-1}{r-6k-n}.$$

3 The game of craps is a gambling game played with two dice in which only the totals count. The player throws the dice and loses at once if the total for the first throw is 2, 3 or 12 and wins at once if it is 7 or 11. Any other total on the first throw is called a "point". If the first throw is a point, the player rolls the dice repeatedly until he either wins by throwing his point again or loses by throwing 7. What is the player's chance to win in at most n throws? Discuss the probability of winning this game.

Let us modify the game by making a player throw the dice n times whether he wins or loses before n throws. The later throws have no effect on the earlier throws and so the probability of winning or losing on the ith throw is unaltered.

To win within n throws, the player can either win by getting a 7 or 11 on the first throw or by getting a point. Since there are 8 ways to get a 7 or 11 and the remaining $n - 1$ throws can be any number at all the player can win in $8(36)^{n-1}$ ways on the first throw.

Let A_i $(i = 2, 3, \ldots, n)$ be the event that the player wins on throw i by throwing the point k. Let $N(A_i)$ denote the number of ways the event A_i can occur. To win on the second throw, the player must get k on the first throw, k on the second throw and any number on the succeeding $n - 2$ throws. Let W_k be the number of ways of getting k; then $N(A_2) = W_k^2(36)^{n-2}$. To win on the third throw he must get k on the first throw, not 7 or k on the second, k on the third throw, and any number on the remaining $n - 3$ throws. Hence $N(A_3) = W_k(30 - W_k) W_k(36)^{n-3}$. Similarly

$$N(A_r) = W_k^2 (30 - W_k)^{r-2} (36)^{n-r}$$

since k must be obtained on the first and r-th throws while neither 7 nor k can be obtained on the intervening $r - 2$ throws and any number is allowed after the rth throw. Therefore the number of ways n_k of winning by getting k is

$$n_k = W_k^2 (36)^{n-2} + W_k^2 (30 - W_k)(36)^{n-3} + W_k^2 (30 - W_k)^2 (36)^{n-4} + \cdots + W_k^2 (30 - W_k)^{n-2}$$

since no two of the events A_i can occur simultaneously. Hence the total number of ways of winning is

$$n_4 + n_5 + n_6 + n_8 + n_9 + n_{10} + 8(36)^{n-1}$$

as the player can win by getting any of the 6 possible points or a seven or 11 on the first throw.

Since there are 36 possible outcomes for each throw the total number of outcomes for n throws is $(36)^n$. Hence the probability p_n of winning is

$$p_n = \frac{n_4}{(36)^n} + \cdots + \frac{n_{10}}{(36)^n} + \frac{8}{36}.$$

To evaluate p_n we need to calculate the following sum:

$$\frac{W_k^2}{(36)^2}\left[1 + \frac{30 - W_k}{36} + \left(\frac{30 - W_k}{36}\right)^2 + \cdots + \left(\frac{30 - W_k}{36}\right)^{n-2}\right]$$

$$= \frac{W_k^2}{(36)^2}\left[\frac{1 - \left(\dfrac{30 - W_k}{36}\right)^{n-1}}{1 - \left(\dfrac{30 - W_k}{36}\right)}\right] = \frac{W_k^2}{36}\left[\frac{1 - \left(\dfrac{30 - W_k}{36}\right)^{n-1}}{6 + W_k}\right].$$

Looking up the values of W_k in the table in problem 2 and using the fact that $n_4 = n_{10}$, $n_5 = n_9$ and $n_6 = n_8$, we have

$$p_n = 2\left[\frac{9}{36}\left(\frac{1 - (27/36)^{n-1}}{9}\right) + \frac{16}{36}\left(\frac{1 - (26/36)^{n-1}}{10}\right)\right.$$
$$\left. + \frac{25}{36}\left(\frac{1 - (25/36)^{n-1}}{11}\right)\right] + \frac{8}{36}$$
$$= \frac{244}{495} - \frac{2}{36}\left[\left(\frac{27}{36}\right)^{n-1} + \left(\frac{16}{10}\right)\left(\frac{26}{36}\right)^{n-1}\right.$$
$$\left. + \left(\frac{25}{11}\right)\left(\frac{25}{36}\right)^{n-1}\right].$$

Theoretically the game could go on forever as the player could repeatedly fail to obtain either a seven or his point on throws 2, 3, 4, ... This is an example in which there are an infinite number of possible outcomes.* The study of such processes is beyond the scope of this book. The interested

* Each outcome can be described by an infinite sequence of numbers $(x_1, x_2, x_3, \cdots)$ where each $x_1 = 2$ or 3 or 4 or $\cdots$ or 12.

reader can consult the references on probability theory listed at the back of this book. However, intuitively it seems true and it can be proven rigorously, that the value of p_n as n becomes extremely large is the probability p of winning at craps. From the fact that x^n $(x < 1)$ becomes extremely small as n becomes extremely large we see that $p = \dfrac{244}{495}$.

Equation (1) states that the probability of winning in n throws is equal to the probability of winning by getting a four plus the probability of winning by getting a five ... plus the probability of winning by getting a 10 plus the probability of winning by getting a 7 or 11 on the first throw. Note that no two of the events "winning by getting an i" $(i = 4, 5, \ldots, 10)$ and "winning by getting a 7 or 11" can occur simultaneously. Moreover if we examine the probability of winning within n throws by getting the point i, $\dfrac{n_i}{(36)^n}$, we see that $\dfrac{n_i}{(36)^n} = \sum_{r=2}^{n}$ (probability of winning on the rth throw by getting i). Again no two of the events "winning on the rth throw by getting i" $(r = 2, \ldots, n)$ can occur simultaneously. These results suggest the following theorem which can be stated more concisely by using the following terminology.

The event A_1 or A_2 or ... or A_n formed from the n events A_i $(i = 1, \ldots, n)$ will be denoted by $A_1 \cup A_2 \cup \cdots \cup A_n$. The events A_i will be said to be *mutually exclusive* if and only if no two of these events can occur simultaneously.

Let $N(A)$ be the number of ways that an event A can occur, that is, the number of elementary events associated with A. The probability of an event A will be denoted by PA.

THEOREM 1 *If the events A_i are mutually exclusive then*

$$P(A_1 \cup A_2 \cup \cdots \cup A_n) = PA_1 + PA_2 + \cdots + PA_n.$$

Proof The corollary to theorem 1.2.2 yields

$$N(A_1 \cup A_2 \cup \cdots \cup A_n) = NA_1 + NA_2 + \cdots + NA_n.$$

Let the total number of elementary events be N. Dividing the last equation by N, we obtain

$$\frac{N(A_1 \cup A_2 \cup \cdots \cup A_n)}{N} = \frac{NA_1}{N} + \frac{NA_2}{N} + \cdots + \frac{NA_n}{N}$$

which yields the desired result upon applying the definition of probability.

It is an instructive exercise to rework problem 3 by using the above theorem. The reader will discover that it is not necessary to introduce the artifical game in order to solve the problem. The following example also illustrates this rule.

4 If six different balls are randomly tossed into 3 different boxes, what is the probability that all boxes will be occupied?

We solve this problem by 3 different methods. The first and second require theorem 1.

(a) Let A_1 be the event that there are 2 balls in each box. Two balls can be placed in each box in only one way. Hence $N(A_1) = \dfrac{6!}{2!\,2!\,2!}$, the number of ways of dividing 6 balls into 3 different groups (the boxes) of 2 each. Since there are 3 boxes, there are 3 choices for each ball and so 3^6 choices for 6 balls (see theorem 3.3.4). Hence $PA_1 = \dfrac{6!\,3^{-6}}{2!\,2!\,2!}$.

Let A_2 be the event that there are 2 balls in one box, 3 balls in another box and 1 ball in the remaining box. Since the boxes are different the number of balls in a particular box is important. There are $3!$ ways that the balls can be placed in the boxes corresponding to the permutations of the numbers 2, 3, and 1. Suppose box 1 has 2 balls, box 2 has 3 balls, and box 3 has 1 ball. We now divide the balls into 3 groups of 2, 3, and 1 balls respectively. This can be done in $\dfrac{6!}{2!\,3!\,1!}$ ways. The same result is obtained for each of the other possible arrangements since the denominator will just contain the numbers $2!$, $3!$, $1!$ in a different order, while the numerator will always be the same. Hence

$$N(A_2) = 3!\,\frac{6!}{2!\,3!\,1!} \quad \text{and} \quad PA_2 = \frac{6!}{2!}\,3^{-6}.$$

Finally let A_3 be the event that there are 4, 1, 1 balls in the three boxes in some order. $N(A_3) = \left(\dfrac{3!}{1!\,2!}\right)\left(\dfrac{6!}{4!\,1!\,1!}\right)$ since the boxes can be occupied in $\dfrac{3!}{1!\,2!}$ different ways and the 6 balls can be divided into 3 groups of 4, 1 and 1 balls in $\dfrac{6!}{4!\,1!\,1!}$ ways. Hence

$$PA_3 = \left(\frac{3!}{2!}\right)\left(\frac{6!}{4!}\right)3^{-6}.$$

Let A be the event that all the boxes will be occupied; then $A = A_1 \cup A_2 \cup A_3$ and the A_i $(i = 1, 2, 3)$ are mutually exclusive. Hence

$$PA = PA_1 + PA_2 + PA_3 = 3^{-6}\left(\frac{6!}{2!\,2!\,2!} + \frac{6!}{2!} + \frac{3!}{2!}\frac{6!}{4!}\right) = \frac{20}{27}.$$

(b) Let x_i $(i = 1, 2, 3)$ represent a box. The sum of the coefficients of $x_1^{i_1}x_2^{i_2}x_3^{i_3}$, where $i_1 + i_2 + i_3 = 6$ and the i's are nonnegative numbers, in

$$(x_1 + x_2 + x_3)^6 \tag{1}$$

is the number of favorable cases. The unordered solutions of the above equation are (2, 2, 2), (2, 3, 1), and (4, 1, 1). Hence the number of favorable outcomes is $\dfrac{6!}{2!\,2!\,2!} + 3!\,\dfrac{6!}{2!\,3!} + 3\,\dfrac{6!}{4!}$, the coefficients, 3! and 3, of the last two terms arise from permutation of (2, 3, 1) and (4, 1, 1). The total number of cases is just the sum of the coefficients in (1). This number, 3^6, is easily obtained by setting $x_1 = x_2 = x_3 = 1$. Once again, we obtain $PA = \frac{20}{27}$.

(c) By theorem 3.3.4 the total number of cases $= 3^6$ and by the corollary to theorem 3.3.5 the number of favorable cases $= \sum_{j=0}^{3}\binom{3}{j}(-1)^j(3-j)^6 = 540$. Hence $PA = \dfrac{540}{27 \cdot 27} = \dfrac{20}{27}$.

Let A be an event; then the event "not A" will be denoted by A'. The following is another useful result which simplifies calculations (cf. theorem 4.1.1).

THEOREM 2 $P(A') = 1 - P(A)$.

Proof The events A and A' are mutually exclusive and either the event A or the event A' must occur. Hence $N(A) + N(A') = N$, the total number of elementary events. Dividing the last equation by N, we obtain $1 = \dfrac{N(A)}{N} + \dfrac{N(A')}{N}$ which is equivalent to the desired result as is proven by the definition of probability.

The application of this result is illustrated in the next problem.

5 What is the probability of getting at least one six on n rolls of a fair die?

Let A be the event we get no sixes in n rolls. Then A' is the event we get at least one six. $N(A) = 5^n$ since we can obtain any of the other 5 numbers on each roll. The total number of cases is 6^n. Hence $PA = \left(\frac{5}{6}\right)^n$ and the required result $P(A') = 1 - \left(\frac{5}{6}\right)^n$.

Of course this problem could have been solved by calculating the number of favorable cases by using theorem 4.1.3. However much more work is involved.

6 Fortunes were once told in the following way. Six long blades of grass were held by one girl with the ends of the blades protruding from the top and bottom of her fist. Another girl would tie the six lower ends in pairs and then the six upper ends in pairs. If the girl succeeded in tying the 6 blades of grass into a ring she would get married within a year. If the blades of grass are tied at random what is the probability that a ring will be formed?

Joining the lower ends in pairs corresponds to dividing the 6 blades into 3 groups of 2 each. Since the groups are indistinguishable this can be accomplished in $\frac{6!}{3!\,(2!)^3} = 15$ ways (cf. discussion following theorem 1.6.1). There are likewise 15 different ways of joining the upper ends. Since the lower and upper ends are joined independently there are $15 \cdot 15 = 225$ equally likely possible outcomes of the experiment.

We now compute the number of favorable outcomes. The lower ends can be connected in 15 ways and for each of these ways it is still possible to form a ring. To show this suppose the 1st blade is tied to the end of the 2nd blade, the 3rd to the 4th, and the 5th to the 6th. In forming a ring it does not matter which of the upper blades we start with. Suppose that the upper end of the second blade is chosen. It can be tied to the upper ends of blades 3, 4, 5 or 6. Suppose it is joined to the 3rd blade; then the upper end of the 4th blade must be joined to the upper end of the 5th or 6th blade. This being done, the remaining blade must be joined to the upper end of the 1st blade. Hence we see that for each of the 15 ways of joining the lower ends there are $4 \cdot 2 \cdot 1 = 8$ ways of joining the lower ends so that a ring will be formed. It follows that the total number of favorable outcomes is $15 \cdot 8 = 120$.

Hence the required probability $= \frac{15 \cdot 8}{15 \cdot 15} = \frac{8}{15}$.

7 Six cards are drawn from an ordinary deck. What is the probability that there will be (a) exactly 1 pair, (b) exactly 2 different pairs, (c) exactly 3 different pairs?

a) We first find the number of 1 pair hands disregarding the suits. There are 13 different pairs to choose from and we can choose 1 pair in $\binom{13}{1}$ ways. This being done, we must now select the 4 different cards from the 12 remaining face values which can be accomplished in $\binom{12}{4}$ ways. Hence a pair and 4 different cards disregarding the suits can be chosen in $\binom{13}{1}\binom{12}{4}$ ways. For each of the 1 pair hands there are $\binom{4}{2}$ ways of choosing suits for the pair of cards since order does not count and 4 ways of choosing a suit for each of the other cards. Hence there are $\binom{4}{2}4^4$ ways of choosing the suits. Hence there are a total of $\binom{4}{2}4^4\binom{13}{1}\binom{12}{4}$ favorable cases.

The total number of cases is just $\binom{52}{6}$, the number of ways of selecting 6 cards from 52; the required probability is therefore

$$\binom{4}{2}4^4\binom{13}{1}\binom{12}{4}\Big/\binom{52}{6}.$$

b) Exactly 2 different pairs (disregarding suits) can be selected in $\binom{13}{2}\binom{11}{2}$ ways since we choose the 2 pairs from 13 possible face values and then choose the 2 different cards from the 11 remaining face values. The suits for the first pair can be selected in $\binom{4}{2}$ ways, the suits for the second pair in $\binom{4}{2}$ ways, while there are 4 choices of suits for each of 2 different cards. Hence for each of the $\binom{13}{2}\binom{11}{2}$ 2-pair hands there are $\binom{4}{2}^2 4^2$ ways of selecting suits and so there are $\binom{13}{2}\binom{11}{2}\binom{4}{2}^2 4^2$ favorable outcomes. Hence the required probability is

$$\binom{13}{2}\binom{11}{2}\binom{4}{2}^2 4^2\Big/\binom{52}{6} = 4\binom{4}{2}^2\frac{13!}{9!}\Big/\binom{52}{6}. \tag{1}$$

The favorable cases in (a) and (b) can be obtained by using theorem 4.2.1. We illustrate its use in (b). Let A_i ($i = 1, 2, \ldots, 13$) be the property that the six cards contain exactly two cards whose face value is i ($i = 11, 12, 13$ represents Jack, Queen, King respectively). $N(A_iA_j) = \binom{4}{2}^2 \binom{44}{2}$ since each pair of i and j cards can be selected in $\binom{4}{2}$ ways; this being done there are 8 cards which cannot be used and so we can select the remaining 2 cards from 44 in $\binom{44}{2}$ ways. Hence $S_2 = \binom{13}{2}\binom{4}{2}^2\binom{44}{2}$ since there are $\binom{13}{2}$ terms of the form $N(A_iA_j)$ in S_2. Similarly, $N(A_iA_jA_k) = \binom{4}{2}^3$ and $S_3 = \binom{13}{3}\binom{4}{2}^3$. Since there are only 6 cards, the S_n with $n > 3$ do not appear. Hence

$$N_{[2]} = S_2 - \binom{3}{2} S_3 = \binom{13}{2}\binom{4}{2}^2\binom{44}{2} - 3\binom{13}{3}\binom{4}{2}^3$$

$$= \binom{4}{2}^2 \frac{13!}{10!\,2!}\left[\frac{44 \cdot 43}{11 \cdot 2} - \frac{18}{3}\right] = \binom{4}{2}^2 \left(\frac{13!}{9!}\right) 4$$

yields the number of hands with *exactly* 2 pairs. Dividing by the total number of outcomes we obtain (1).

(c) Exactly 3 different pairs (disregarding suit) can be chosen in $\binom{13}{3}$ ways. The suit, for each of these 3 different pairs, can be selected in $\binom{4}{2}$ ways. Hence the total number of favorable outcomes is $\binom{13}{3}\binom{4}{2}^3$ yielding a probability of $\binom{13}{3}\binom{4}{2}^3 \Big/ \binom{52}{6}$.

8 a) Guess the least number of persons required in order that the probability exceeds $\frac{1}{2}$ that two or more of them have the same birthday (year of birth need not match)? Then calculate the number.

b) What is the least number of strangers required so that the probability is at least $\frac{1}{2}$ that one of their birthdays matches yours?

(a) We simplify the problem by ignoring February 29 as a possible birthday and assuming that the other 365 days are equally likely birthdates.

Let r be the number of people. Since there are 365 possibilities for the birthday of each person there are $(365)^r$ possible outcomes. The number of ways for there to be no matching birthdays is $365 \cdot 364 \cdots (365 - r + 1)$.

Hence the probability of there being no matches is $\frac{{}_{365}P_r}{(365)^r}$. Applying theorem 2, we obtain the probability of at least one match:

$$1 - \frac{365!}{(365 - r)!\,(365)^r}.$$

A table of logarithms shows that for $r = 22$, the probability is 0.476 while for $r = 23$, the probability for at least one success is 0.507. Hence $r = 23$ is the least number of people required so that the probability is slightly greater than $\frac{1}{2}$. For 60 people, the probability is 0.994; it would be an extremely good wager to bet that 2 or more people have the same birthday.

This problem can be interpreted as placing r balls in $n = 365$ cells (c.f. theorem 3.3.4).

(b) Again we let r be the number of people and we have $(365)^r$ possible outcomes. If none of the r people have the same birthday as you their birthdays must fall on the remaining 364 days. Hence there are $(364)^r$ ways for r people to miss matching your birthday. Therefore the probability that none of them has your birthday is $\left(\frac{364}{365}\right)^r$; and so the probability that at least one matches is $1 - \left(\frac{364}{365}\right)^r$. Solving for the least value of r such that $1 - \left(\frac{364}{365}\right)^r \geqslant \frac{1}{2}$ (by using logarithms) we find that $r = 253$.

9 A particle at the origin is displaced along a straight line by a series of steps of equal length, say unity. Each step is independent of all previous steps and a step is taken in the forward direction with probability p and in the backward direction with probability q $(p + q = 1)$. After taking s such steps, the particle could be at any of the points $-s, -s + 1, \ldots, -1, 0, 1, \ldots, s - 1, s$. What is the probability, $P(r, s)$, that the particle arrives at the point r after s steps $(s \geqslant |r|)$?

We assume that $p = \frac{k}{m}$ and $q = \frac{m - k}{m}$ are rational numbers*, in other words k and m are positive integers. If not, we can approximate p and q by a sequence of rational numbers.

---

* An advanced treatment appears in [9].

The random walk executed by the particle can actually be realized in the following manner. Get a deck of m cards and mark them with the numbers $1, 2, \ldots, m$. Shuffle the deck thoroughly and draw a card. If the card is marked $1, 2, \ldots, k$ the particle is displaced by $+1$; otherwise it is displaced by -1. Return the card to the deck and repeat the procedure s times. It is not difficult to see that the probability that the particle is displaced forward is $p = \dfrac{k}{m}$ while the probability that the particle is displaced backwards is $q = \dfrac{m - k}{m}$.

Each sequence of s steps will be uniquely represented by a sequence of numbers $x_1, x_2, \ldots, x_s$ where each $x_i = 1, 2, \ldots, m$ and conversely. The particle will be at the point r if and only if the particle has taken s_1 positive steps and s_2 negative steps with $r = s_1 - s_2$. Since $s = s_1 + s_2$, we see that $s_1 = \dfrac{r + s}{2}$ while $s_2 = \dfrac{s - r}{2}$. The particle will move s_1 positive steps and s_2 negative steps if and only if s_1 of the x_i's in the above sequence take values from $1, 2, \ldots, k$ while s_1 of the x_i's take values from $k + 1, k + 2, \ldots, m$. Hence the numbers of favorable cases is just the same as the number of sequences which have s_1 places filled by the numbers $1, 2, \ldots, k$ and s_2 places filled by the numbers $k + 1, k + 2, \ldots, m$. The number of ways in which the places filled by the numbers $1, 2, \ldots, k$ can be selected is $\dbinom{s}{s_1}$, since for each of these ways the places filled by the other s_2 numbers are uniquely determined. For each such selection we can form $k^{s_1} (m - k)^{s_2}$ different sequences since each of the fixed s_1 places can be filled in k ways and the s_2 places in $m - k$ ways. Hence the total number of sequences of the required form is $\dbinom{s}{s_1} k^{s_1} (m - k)^{s_2}$. The total number of outcomes is just the same as all possible sequences of length s which can be formed by letting each $x_i = 1, 2, \ldots, m$. There are m^s such sequences. Hence

$$P(r, s) = \binom{s}{s_1} k^{s_1} (m - k)^{s_2}/m^s = \binom{s}{s_1} \left(\frac{k}{m}\right)^{s_1} \left(\frac{m - k}{m}\right)^{s_2}$$

$$= \binom{s}{s_1} p^{s_1} q^{s_2} = \binom{s}{\frac{r + s}{2}} p^{\frac{r+s}{2}} q^{\frac{s-r}{2}}. \tag{1}$$

10 A drunkard takes s random steps of equal length along a straight road. At any point his probability is $p = \frac{1}{2}$ of taking a forward step and $q = \frac{1}{2}$ of taking a backward step. There is a wall on the road located at r_1 steps in the forward direction. Each time the drunkard walks into the wall he is bounced 1 step backwards.* Find the probability $P(r, s; r_1)$ that the drunkard will be at r steps from his starting position ($r \leqslant r_1 \leqslant s$).

We represent the motion of the drunkard on the xt plane as follows. The distance of the drunkard from the origin (his starting point) is represented by distance along the x-axis while the number of steps is represented by the distance along the t-axis. The same scale of measurement is used on both axes, the unit length being the length of a step. If the drunkard has reached the position x after t steps we represent this fact by plotting the point (x, t). Hence the position of the drunkard at each of the s steps of his journey is represented by the points $(0, 0)$, $(x_1, 1)$, $(x_2, 2)$, ... (x_s, s). Connect $(0, 0)$ to $(x_1, 1)$ by a straight line segment and (x_i, i) to $(x_{i+1}, i + 1)$ by a straight line segment. In this way the path of the drunkard will be represented by a polygonal line. The figure below shows two possible paths of the drunkard: *OFBEC* and *OFGHIC*. For $x < r$, the point on the polygonal line moves one unit upward for each step of the drunkard and one unit to the right or left (with probability $\frac{1}{2}$ in each case). If $x = r_1$ the point moves one unit to the left with probability 1.

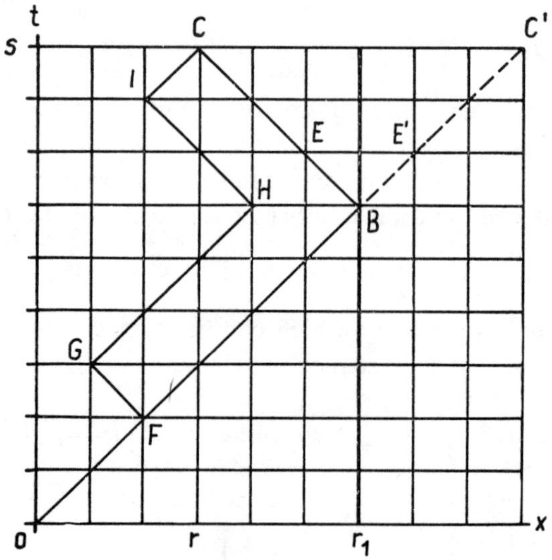

* That is, on the next step the probability is 1 that he will step backwards.

Using the interpretation at the beginning of the solution to problem 9, it is not difficult to show that if C is reached in s steps the probability of traversing the path OBC is $\frac{1}{2^{s-1}}$. In fact we can show more generally that if the path meets the barrier k times in s steps and does not terminate at the barrier, the probability of traversing the path is $\frac{1}{2^{s-k}}$. In this example, our deck consists of 2 cards marked L (for left) and R (for right). We draw s cards from the deck, and so obtain a sequence $(x_1, x_2, \ldots, x_s)$ of L's and R's. We move a particle on the line in accordance with this sequence, thereby simulating the drunkard's walk. However if the barrier is reached at step i then we ignore x_{i+1} and always move the particle to the left. Suppose the barrier is reached at steps $i_1 - 1, i_2 - 1, \ldots, i_k - 1$; then the values of $x_{i_1}, \ldots, x_{i_k}$ are immaterial, that is, any two sequences $(x_1, x_2, \ldots, x_s)$ and $(y_1, y_2, \ldots, y_s)$ with $x_i = y_i$ except for $i = i_1, i_2, \ldots, i_k$ will yield the same path. Hence the number of favorable sequences for the path is 2^k since we have 2 choices L and R for each $x_{i_1}, \ldots, x_{i_k}$. Since the total number of sequences is 2^s, the required probability is $\frac{2^k}{2^s} = \frac{1}{2^{s-k}}$. In particular, if $k = 0$, the probability of traversing such a path (e.g. $OFGHIC$) is $\frac{1}{2^s}$.

Instead of counting a path that reaches C after k reflections 2^k times, we proceed as follows: Consider first a path like OBC which has undergone just one reflection at r_1. By reflecting this path about the vertical line through r_1 we obtain a trajectory leading to the image point C' with coordinates $(2r_1 - r, s)$. Conversely, for every path leading to the image point C', having crossed the line $x = r_1$ once, there is exactly one path which leads to C after a single reflection. Hence, instead of counting twice each trajectory reflected once, we can add a uniquely defined trajectory leading to C'. Consider next a path like $OABDC$ which leads to C after 2 reflections (see diagram below).

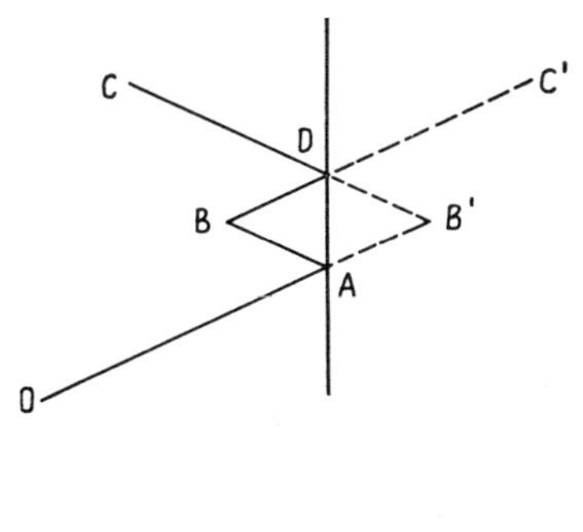

A path like this should be counted 2^2 times. If the barrier is removed, there are 2 paths $OAB'DC'$ and $OABDC'$ leading to the image point and a third, $OAB'DC$, which leads to C. These 3 extra paths along with $OABDC$ give exactly 4 paths leading either to C or C' in the absence of the reflecting barrier. The argument can be generalized to the case of k reflections by using induction.

Hence in order for the moving point in the xt-plane to reach C with reflections it is both necessary and sufficient that it reachs either the point C or C' while moving without a reflecting barrier. The probability of reaching C was computed in the previous problem and is

$$P(r, s) = \binom{s}{\frac{r+s}{2}} 2^{-s} \tag{1}$$

and similarly, the probability of reaching C' is

$$P(2r_1 - r, s) = \binom{s}{r_1 + \frac{s-r}{2}} 2^{-s}. \tag{2}$$

The required probability is therefore given by

$$P(r, s; r_1) = P(r, s) + P(2r_1 - r, s).$$

11 In problem 10, suppose that instead of a wall there is some muddy ground and the drunkard becomes stuck and cannot move if he steps on it. Find the probability, $P(r, s; r_1)$, under this hypothesis.*

Again we consider all possible sequences $(x_1, \ldots, x_s)$ of L's and R's which lead to C if the absorbing barrier at $x = r_1$ were not there. However, we must exclude all sequences in which the corresponding paths meet the line $x = r_1$ to find the favorable cases. Such a forbidden sequence uniquely defines another sequence leading to C' and conversely (see diagram below). To show this, reflect about the line $x = r_1$ the part of the forbidden path above its last point of contact with the line $x = r_1$ before arriving at C. We obtain a path leading to the image point C'. Conversely, for every path leading to C' we obtain by reflection a forbidden path leading to C, since every

* Problem 10 is called a random walk with a reflecting barrier while problem 11 is known as a random walk with an absorbing barrier. These are simple prototypes of actual physical problems in diffusion.

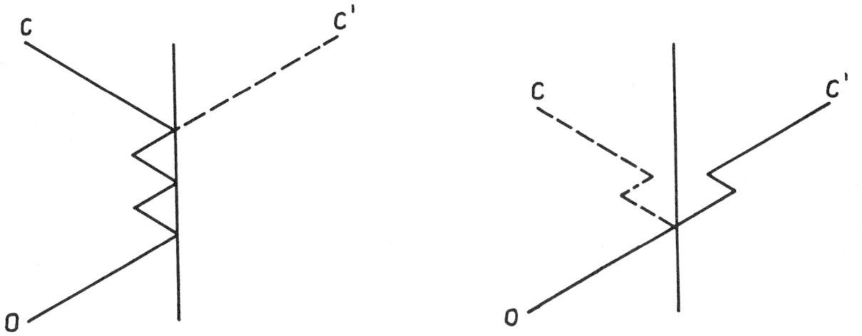

path leading to C' must cross the line $x = r_1$. Hence the number of favorable sequences equals the number of sequences yielding a path leading to C with no barrier minus the number of sequences yielding a path leading to C' with no barrier. Dividing this last result by the total number of sequences, 2^s, we obtain

$$P(r, s; r_1) = P(r, s) - P(2r_1 - r, s)$$

where the probabilities appearing on the right hand side are defined in (1) and (2) of problem 11.

12 Bridge is a game played with an ordinary pack of 52 cards. There are 4 players known as N (north), S (south), E (east), and W (west). N and S are partners, as are E and W. The only information we require about the game is that the pack is well-shuffled and 13 cards are dealt to each player.* A set of 13 cards held by a particular player is called a hand; the order in which the cards are arranged does not matter. The distribution into 4 groups of 13 cards each is called a *deal*.

a) Show that no matter how the cards are dealt, as long as each player gets 13 cards, there are $\binom{52}{13}$ possible hands.

b) Prove that the probability of W having k aces is the same as the probability that an arbitrary hand of 13 cards contains k aces.

c) Verify that (i) the probability that N receives m and S receives n spades is the same as (ii) the probability that when 2 hands of 13 cards each are drawn at random from an ordinary deck of cards, the first contains m and the second contains n spades.

---

* In an ordinary game, the deck is not thoroughly shuffled and the cards are not arranged at random. However we adopt this assumption.

a) The total number of different deals is the number of ways that 52 cards can be divided into 4 different groups of 13 cards each. The number of ways is $\dfrac{52!}{13!\,13!\,13!\,13!}$ by theorem 1.6.1. Let x be the number of different hands a particular player can hold. Each hand gives rise to $\dfrac{39!}{13!\,13!\,13!}$ different deals, since the remaining 39 cards can be distributed in this many ways to the other 3 players. Hence $\dfrac{39!}{(13!)^3}\,x = \dfrac{52!}{(13!)^4}$, the total number of deals, and so $x = \dbinom{52}{13}$.

This result is intuitively obvious; since the deck is well-shuffled the player might just as well choose any 13 cards from the deck. The remaining parts of the problem are also intuitively clear.

b) 13 cards can contain exactly k aces in $\dbinom{4}{k}\dbinom{48}{13-k}$ ways. This is because the aces can be chosen in $\dbinom{4}{k}$ ways and for each of these ways the remaining $13 - k$ cards, being non-aces, can be selected in $\dbinom{48}{13-k}$ ways. The total number of 13 card hands is $\dbinom{52}{13}$ and so the required probability is $\dbinom{4}{k}\dbinom{48}{13-k}\Big/\dbinom{52}{13}$.

We compare this result to the probability that W's hand contains k aces. As above, W's hand can be selected in $\dbinom{4}{k}\dbinom{48}{13-k}$ ways. For each of these ways there remain 39 cards to be evenly divided among N, E, and S. This can be done in $\dfrac{39!}{(13!)^3}$ ways. Hence the number of favorable outcomes $= \dbinom{4}{k}\dbinom{48}{13-k}\dfrac{39!}{(13!)^3}$. The total number of deals is $\dfrac{52!}{(13!)^4}$ and so the required probability is

$$\binom{4}{k}\binom{48}{13-k}\frac{39!}{(13!)^3}\Big/\frac{52!}{(13!)^4} = \binom{4}{k}\binom{48}{13-k}\Big/\binom{52}{13}.$$

c) (ii) We first find the number of ways that a hand of 13 cards drawn from an ordinary deck contains m spades and another 13 cards drawn from the remainder contains n spades. m spades can be selected from 13 spades in $\binom{13}{m}$ ways and for each of these ways we can select $13 - m$ non-spades from 39 non-spades in $\binom{39}{13-m}$ ways. Therefore the number of ways that the first thirteen cards contains m spades is $\binom{13}{m}\binom{39}{13-m}$. For each way in which the first 13 cards were selected we can select n spades from the $13 - m$ remaining spades in $\binom{13-m}{n}$ ways and the remaining $13 - n$ non-spade cards from the $39 - (13 - m)$ non-spades in $\binom{26+m}{13-n}$ ways. Hence the number of favorable cases is

$$\binom{13}{m}\binom{39}{13-m}\binom{13-m}{n}\binom{26+m}{13-n}. \tag{1}$$

The first 13 cards can be selected in $\binom{52}{13}$ ways without restriction and for each of these ways the second 13 cards can be selected in $\binom{39}{13}$ ways yielding a total of $\binom{52}{13}\binom{39}{13}$ equally likely possible outcomes. Therefore the required probability is

$$\binom{13}{m}\binom{39}{13-m}\binom{13-m}{n}\binom{26+m}{13-n}\bigg/\binom{52}{13}\binom{39}{13}. \tag{2}$$

(i) It is intuitively clear that the method of distributing cards does not effect the probabilities provided that the deck is well-shuffled. We assume that we give 13 cards at a time to each player in the order N, S, W, E. The total number of ways that N has m spades and S has n spades in the 26 distributed cards is given by (1) above. For each of these ways there are $\binom{26}{13}$ ways of distributing 13 cards to W and $\binom{13}{13}$ ways of distributing the remaining cards to E. Hence the total number of favorable deals is

$$\binom{13}{m}\binom{39}{13-m}\binom{13-m}{n}\binom{26+m}{13-n}\binom{26}{13}\binom{13}{13}$$

Dividing this result by the total number of deals $\dfrac{52!}{(13!)^4}$ we obtain (2).

13 a) What is the probability that N receives all 13 spades?

b) What is the probability that N has a perfect bridge hand (13 of one suit)?

a) There is 1 way of getting 13 spades and $\binom{52}{13}$ possible outcomes. Hence the required probability is $1\Big/\binom{52}{13}$.

b) Since there are 4 suits, we obtain $4\Big/\binom{52}{13}$. This is an extremely small number; there is about 1 chance in 160 billion of getting a perfect hand on a single deal.

14 a) What is the probability that N is dealt a hand at bridge which contains no hearts?

b) Suppose S picks up his hand and notes that he has 5 hearts. What is the probability, as far as S is concerned, that N has no hearts?

c) Suppose E bids "one heart" which means E has precisely 3 hearts. What is the probability, as far as S is concerned, that N has no hearts?

a) Since N has no hearts, his hand of 13 cards must be comprised from the 39 non-hearts and so N can have $\binom{39}{13}$ possible hands. Hence the required probability is $\binom{39}{13}\Big/\binom{52}{13}$.

b) There are 39 cards from which N's hand of 13 can be selected since S has 13 cards. Hence there are $\binom{39}{13}$ possible hands for N as far as S is concerned. Among these 39 cards, there are only 8 hearts since S has 5. Excluding these hearts, there are 31 non-hearts from which 13 cards must be selected in order for N to have a hand without hearts. This can be done in $\binom{31}{13}$ ways and so the required probability is $\binom{31}{13}\Big/\binom{39}{13}$.

c) There are now 26 cards from which N's hand of 13 can be selected since S and E have 26 cards. Hence there are $\binom{26}{13}$ possible hands for N. There are 8 hearts missing from these 26 cards since E has 3 and S has 5.

Excluding the remaining 5 hearts, there are $\binom{21}{13}$ possible hands for N which are void of hearts. Hence the required probability is $\binom{21}{13}\Big/\binom{26}{13}$.

15 What is the probability that a randomly dealt hand at bridge* will be void in at least one suit?

We use theorem 4.1.3. Let A_i be the property that a randomly dealt hand contains no cards in the ith suit ($i = 1, 2, 3, 4$). Then $N(A_i) = \binom{39}{13}$ since then are $\binom{39}{13}$ cards which are not of suit i and we select 13 of these; hence $S_1 = 4\binom{39}{13}$. To be void in 2 suits, 26 cards must be excluded and 13 chosen from the remaining 26. This can be done in $\binom{26}{13}$ ways and so $S_2 = \binom{4}{2}\binom{26}{13}$. Since $N(A_iA_jA_k) = \binom{13}{13}$, it follows that $S_3 = \binom{4}{3}$. Hence $N(A_1 \cup A_2 \cup A_3 \cup A_4) = 4\binom{39}{13} - 6\binom{26}{13} + 4$ and so the required probability is

$$\left[4\binom{39}{13} - 6\binom{26}{13} + 4\right]\Big/\binom{52}{13}.$$

16 Five-card draw poker is played with an ordinary deck of 52 cards. The cards are well-shuffled and 5 cards are dealt to each of 2 or more players. The five cards dealt to a player are a hand. The hands are ranked in the following order (the lower the number the better the hand).

i) A *straight flush* is a hand in which the cards form a sequence such as 1 2 3 4 5 or 10 J Q K 1 where all cards are of the same suit.

ii) *Four of a kind* is a hand with 4 cards of equal face value e.g. 111K.

iii) A *full house* is a hand with 1 pair and 1 triple of cards with equal face value such as 2 2 4 4 4.

iv) A *flush* is a hand of one suit.

(v) A *straight* is a hand in sequence without regard to suit.

vi) *Three of a kind* is a hand with 3 equal face values and 2 other different cards e.g. 1 1 1 J K.

---

* For further results consult E.Borel and A.Cheron, *Théorie Mathématique du Bridge*, Gauthier-Villars, Paris (1955).

vii) Two *pairs* is a hand with 2 pairs of equal face value plus one other card e.g. 1 1 J J 2.

viii) One *pair* is a hand with 1 pair of equal face value and 3 different cards e.g. 1 1 2 5 7.

ix) *Nothing* is a hand which is none of the preceding.

Find the probabilities of being dealt the above type of hands. Is the method of ranking justified?

Since the deck is well-shuffled getting a hand in poker is equivalent to drawing 5 cards from a deck of 52 cards. Hence the total number of possible hands is $\binom{52}{5}$. To solve the problem, we need only find the favorable cases in each of the parts above and divide by $\binom{52}{2}$. Hence, we merely list the favorable outcomes.

i) There are 10 straights in each suit, namely 1 2 3 4 5, 2 3 4 5 6, ..., 10 J Q K 1. Since there are 4 suits there are 40 possible straight flushes.

ii) There are 13 choices for the 4 repeated cards and then 48 choices for the remaining card yielding $13 \cdot 48 = 624$ different outcomes.

iii) There are 13 choices for the 1 pair and then 12 choices for the triple. For given face values (e.g. suppose we have selected 1 1 2 2 2) we have a choice as to the suits; for the pair we have $\binom{4}{2}$ choices while for the triple we have $\binom{4}{3}$ choices. Hence the total number of full houses is $13 \cdot 12 \cdot 4 \cdot 6 = 3744$.

iv) There are $\binom{13}{5}$ hands of a particular suit and so $4\binom{13}{5} = 5148$ flushes.

v) There are 10 straights (not counting the suit), and 4 choices for each of the five cards in a given sequence. Hence each sequence of 5 consecutive numbers yields 4^5 straights and so there are $10 \cdot 4^5 = 10{,}240$ straights.

vi) The 3 of a kind (as regard to face value) can be chosen in 13 ways and for each of these ways the suits can be chosen in $\binom{4}{3} = 4$ ways. The 2 different cards can be chosen in $\binom{12}{2}$ ways as regards face values and their suits specified in 4^2 ways. Hence there are $(13)\binom{12}{2}4^3$ hands which are 3 of a kind.

vii) By reasoning as in problem 7(b) we see that there are $\binom{13}{2} 11 \cdot 6 \cdot 6 \cdot 4$ hands of this type.

viii) By reasoning as in problem 7(a) we see that there are $13 \binom{12}{3} 6 \cdot 4^3$ hands of this type.

ix) Disregarding suits there are $\binom{13}{5}$ hands consisting of cards whose face values are all different. Since there are 10 straigths, there are $\binom{13}{5} - 10$ hands with different face values which are not straights. For each of these hands, 4 suits may be chosen for each card and so for fixed face values there are 4^5 different hands when suits are taken into account. However, 4 of these would be flushes and so there are $4^5 - 4$ hands which are not flushes for given face values. Hence there are

$$\left[\binom{13}{5} - 10\right] [4^5 - 4]$$

nothing hands.

17 (The problem of the n liars) The n people A_i ($i = 1, \ldots, n$) each tell the truth with probability q ($0 < q < 1$, q rational). Suppose that A_1 makes a statement; then A_n says that A_{n-1} says that A_{n-2} says that ... says that A_2 says that A_1 was telling the truth. What is the probability that A_1 was actually telling the truth?

The above problem can be formulated as follows: A slip marked with T is given to A_1, who passes it to A_2, who passes it to A_3 and so on; finally A_n passes it to a judge (a truthful person). At the ith stage, A_i shuffles a deck of cards numbered $1, 2, \ldots, m + r$ and draws a card. If he draws a card numbered $1, \ldots, m$ he changes the letter on the slip of paper (T is changed to L and L is changed to T); otherwise he does not change it. Here $p = \dfrac{m}{m + r}$ is the probability of changing the letter and $q = \dfrac{r}{m + r}$ is the probability of leaving the letter unaltered. Suppose the judge observes a T at the end of the process. What is the probability that A_1 left the letter unchanged (told the truth)?

We first find the total number of outcomes. In order for the final letter to be T, it must have been changed an even number of times, say $2k$, where

$0 \leqslant 2k \leqslant n$. There are $\binom{n}{2k}$ ways to choose which people make the changes. Suppose these people have been chosen. The number of ways that they can change the letter, while the remaining $n - 2k$ people will leave it alone is

$$\underbrace{m\,m\cdots m}_{2k}\,\underbrace{r\,r\cdots r}_{n-2k} = m^{2k}r^{n-2k}.$$

Hence the number of ways the final letter can be T, given that there were $2k$ changes, is $\binom{n}{2k} m^{2k}r^{n-2k}$. Therefore the total numbers of outcomes is

$$\binom{n}{0} r^n + \binom{n}{2} m^2 r^{n-2} + \binom{n}{4} m^4 r^{n-4} + \cdots, \tag{1}$$

where the series terminates as soon as $2k > n$. To evaluate (1) in closed form, apply the binomial theorem to $(r + m)^n$ and $(r - m)^n$. We obtain

$$(r + m)^n = \binom{n}{0} r^n + \binom{n}{1} r^{n-1}m + \binom{n}{2} r^{n-2}m^2 + \binom{n}{3} r^{n-3}m^3 + \cdots + \binom{n}{n} m^n,$$

$$(r - m)^n = \binom{n}{0} r^n - \binom{n}{1} r^{n-1}m + \binom{n}{2} r^{n-2}m^2 + \binom{n}{3} r^{n-3}m^3 + \cdots + (-1)^n \binom{n}{n} m^n.$$

Adding these two equations and dividing by 2, the right side reduces to (1) and so the total number of outcomes is

$$\frac{(r + m)^n + (r - m)^n}{2}. \tag{1'}$$

For a favorable situation an even number of people must change the letter, but A_1 must not be one of them. Hence if $2k$ people change the letter, they can be chosen in $\binom{n-1}{2k}$ ways. As before, the number of ways that $2k$ people can change the letter while the other $n - 2k$ people leave it alone is $m^{2k}r^{n-2k}$.

Therefore the total number of favorable outcomes is

$$\binom{n-1}{0} r^n + \binom{n-1}{2} m^2 r^{n-2} + \binom{n-1}{4} m^4 r^{n-4} + \cdots$$

$$= r\left[\binom{n-1}{0} r^{n-1} + \binom{n-1}{2} m^2 r^{n-3} + \binom{n-1}{4} m^4 r^{n-5} + \cdots\right]$$

$$= r\left[\frac{(r+m)^{n-1} + (r-m)^{n-1}}{2}\right].$$

Dividing the above result by (1′) we obtain the required probability

$$r\left[\frac{(r+m)^{n-1} + (r-m)^{n-1}}{(r+m)^n + (r-m)^n}\right]$$

$$= \frac{r}{r+m}\left[\frac{1 + (r(r+m)^{-1} - m(r+m)^{-1})^{n-1}}{1 + (r(r+m)^{-1} - m(r+m)^{-1})^n}\right]$$

$$= q\left[\frac{1 + (q-p)^{n-1}}{1 + (q-p)^n}\right].$$

18 a) (Banach's matchbox problem*) A mathematican always carries two boxes of matches. Every time he has to light a match, he selects at random one box or the other. When a box is empty he absentmindedly puts it back in his pocket and only discovers that it is empty when he wants to remove another match. Suppose each box originally contains n matches. What is the probability that there are k $(0 \leqslant k \leqslant n)$ left in the other box when he first finds an empty box?

b) Evaluate the sum

$$\binom{2n}{n} + 2\binom{2n-1}{n} + 2^2\binom{2n-2}{n} + \cdots + 2^n\binom{n}{n}.$$

a) This problem is equivalent to finding the probability that when the mathematician tries to take an $(n+1)$st match from the empty box E, exactly $n-k$ have been taken from the other box. Formulated in this manner we see that there is no harm in assuming that there are an infinite number of matches in each box to begin with. The advantage of having an infinite number of matches is that we can consider it equally likely that a match

---

* Stephen Banach (1892–1945), a Polish mathematician.

is selected from either box. The number of ways that $n + (n - k)$ matches can be selected is 2^{2n-k}. This is the total number of outcomes since the $(2n - k + 1)$st match is assumed to come from E and so there is no choice for it. The favorable outcomes are those in which n matches are chosen from E and the remaining $n - k$ from the other box. This can be done in $\binom{2n-k}{n}$ ways; and so the desired probability is $\binom{2n-k}{n} 2^{k-2n}$.

b) Let A_i be the event that when the mathematician discovers that he picked an empty box there are i matches in the other box. Then the event $A_0 \cup A_1 \cup A_2 \cup \cdots \cup A_n$ is bound to happen; and so $P(A_0 \cup A_1 \cup \cdots \cup A_n) = 1$. Since the events A_i are mutually exclusive,

$$1 = P(A_0 \cup A_1 \cup \cdots \cup A_n) = PA_0 + PA_1 + \cdots + PA_n = \binom{2n}{n} 2^{-2n} + \binom{2n-1}{n} 2^{-2n+1} + \cdots + \binom{n}{n} 2^{-n}.$$

Multiplying by 2^{2n}, we obtain

$$\binom{2n}{n} + 2\binom{2n-1}{n} + 2^2\binom{2n-2}{n} + \cdots + \binom{n}{n} 2^n = 2^{2n}.$$

19 Find the probability that 2 players are even after having played $2n$ games of heads and tails.

Since each toss can result in heads or tails there are 2^{2n} possible outcomes in $2n$ games. A tie will result if and only if there have been n heads and n tails. Since the n heads can be selected in $\binom{2n}{n}$ ways, the required probability is $2^{-2n}\binom{2n}{n} = \dfrac{(2n)!}{2^{2n}(n!)^2}$. Using Stirling's formula, we obtain the following approximation foı large n

$$\frac{(2n)^{2n} e^{-2n} \sqrt{4\pi n}}{2^{2n} n^{2n} e^{-2n} 2\pi n} = \frac{1}{\sqrt{\pi n}}.$$

Hence, the probability of a tie decreases when the number of games, fixed in advance, is increased; for 200 games the probability is approximately $\frac{1}{25}$ and for 2,000,000 games it is approximately $\frac{1}{2500}$. Long runs favoring one player are improbable, but they will happen if the games last sufficiently long. Hence there will be long periods of time during which the players are not even*.

20 A child removes all the labels from n different kinds of baby food. The mother feeds the child 1 jar a day for the the next n days, guessing the contents of the jar. What is the probability that she was wrong each time? Guess what the probability is for large values of n, say 1,000,000 jars.

The total number of cases is $n!$ By example 4.2.1 the number of favorable outcomes is $n!\left(1 - 1 + \frac{1}{2!} - \frac{1}{3!} + \cdots + (-1)^n \frac{1}{n!}\right)$. Hence the required probability is $\left(1 - 1 + \frac{1}{2!} + \cdots + (-1)^n \frac{1}{n!}\right)$. For the large values of n the above probability is almost e^{-1} (cf. equation 3.2.5).

21 a) $f + t$ people are queued up at a theater box office; f of them have only five-dollar bills and the other t have only ten-dollar bills. There is no cash in the box office when it opens and each patron in turn is going to buy a single five-dollar ticket. What is the probability that none of them will have to wait for change?

b) Assume that there were initially n five-dollar bills in the till. What is the probability that one of them will have to wait for change?

a) The total number of outcomes is just all possible arrangements of $f + t$ people, f having five-dollar bills and the rest ten-dollar bills; this is $\binom{f+t}{f}$.

We can represent these possibilities graphically in the xy-plane; a unit on the x-axis represents \$5 and a unit on the y-axis represents \$10. This geometrical viewpoint will help us determine the favorable outcomes. Starting from the point $A_0 = (0, 0)$ draw a segment of length 1 either to the right (if the first customer has a five-dollar bill or upwards (if he has only a ten-dollar bill). From the end point A_1 of this segment, draw a segment A_1A_2 of length 1 either to the right or upwards according to whether the second customer has a five-dollar bill or a ten-dollar bill. In general, from the point A_i lay off a segment A_iA_{i+1} of length 1 horizontally or vertically according to whether the $(i + 1)$st customer has a 5 or 10-dollar bill ($i = 0, 1, 2, \ldots, f+t-1$). In this way, each of the $\binom{f+t}{f}$ possible arrangements of the $f + t$ customers uniquely corresponds to a path $A_0A_1A_2 \cdots A_{f+t}$ con-

* For further details see Chapter 3 of W. Feller, *Introduction to Probability Theory*, vol. 1, John Wiley and Sons, New York (1960).

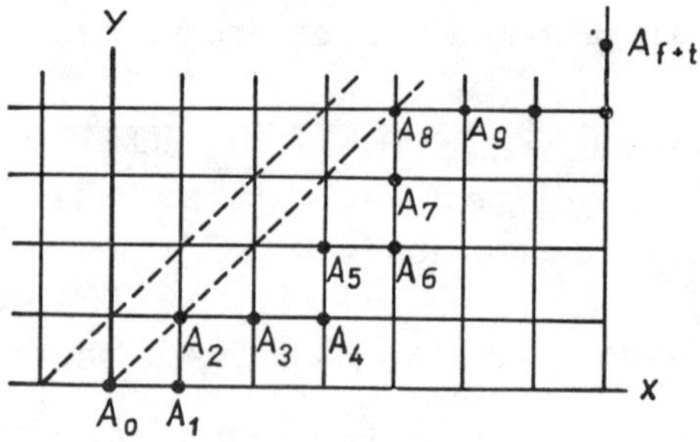

sisting of f horizontal and t vertical segments and conversely (see diagram below). All of these paths end at the point $A_{f+t} = (f, t)$.

We now find a geometrical interpretation of the fact that no customer has to wait for change. This occurs if and only if at any point of such a line there are as many people with five-dollar bills as with ten-dollar bills ahead of this point (the point itself being included). Geometrically, this means that for each $A_i = (x, y)$ we must have $x \geqslant y$. In other words, every favorable outcome corresponds to a path $A_0A_1A_2 \cdots A_{f+t}$ which lies entirely below or on the straight line $L : y = x$. The first segment of such a path must be horizontal.

Hence every path corresponding to an unfavorable outcome must cross the line L or equivalently have a vertex on the $L_1 : y - x = 1$. It is easier to find the number of such paths and to subtract this number of paths, then to calculate the favorable paths directly. For $t > f$, the probability of no one waiting is zero* and so we are only interested in the case $t \leqslant f$. Let $A_0A_1 \cdots A_{k-1}A_k \cdots A_{f+t}$ be a path with a vertex on L_1 and let A_k be the *first* of its vertices which lies on the line L_1 (see diagram below). Reflect the

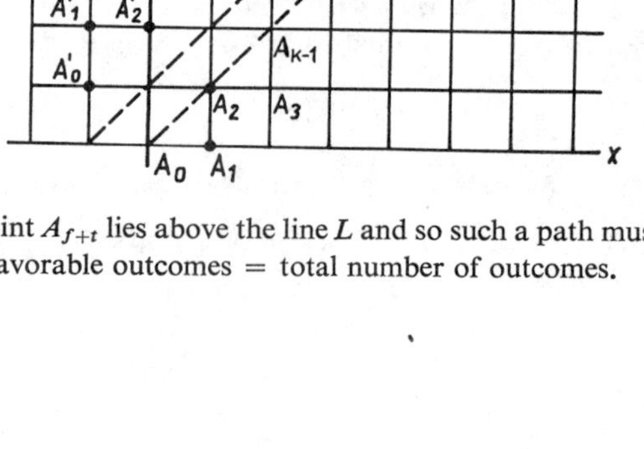

* If $t > f$ the point A_{f+t} lies above the line L and so such a path must intersect L_1. Here the number of unfavorable outcomes = total number of outcomes.

portion $A_0A_1 \cdots A_{k-1}A_k$ of this path using the line L_1 as a mirror. We obtain a path $A'_0A'_1 \cdots A'_{k-1} \cdots A_{f+t}$ which joins $A'_0 = (-1, 1)$ to A_{f+t}. Furthermore, for $f \geqslant t$ every path joining the point A'_0 and A_{f+t} necessarily intersects L_1. If A_k is the *first* vertex of $A'_0A'_1 \cdots A'_{k-1}A_k \cdots A_{f+t}$ on the line L_1 then by reflecting $A'_0A'_1 \cdots A'_{k-1}A_k$ we obtain a path $A_0A_1 \cdots A_{k-1}A_k \cdots A_{f+t}$ joining A_0 to A_{f+t} and having a vertex on L_1. Hence for $t \leqslant f$ the number of paths joining A_0 to A_{f+t} and having a vertex on the line L_1 coincides with the number of paths joining A'_0 to A_{f+t}. A path joining $A_0 = (-1, 1)$ to $A_{f+t} = (f, t)$ consists of $f + 1$ horizontal segments and $t - 1$ vertical segments; hence there are $\binom{f+t}{t-1}$ of them.

Since there are $\binom{f+t}{f}$ outcomes and $\binom{f+t}{t-1}$ unfavorable outcomes the number of favorable outcomes is

$$\binom{f+t}{f} - \binom{f+t}{t-1} = \frac{(f+t)!}{f!\,(t-1)!}\left[\frac{1}{t} - \frac{1}{f+1}\right].$$

Hence the required probability p is

$$\begin{aligned} p &= \frac{(f+t)!\,(f-t+1)}{(f+1)!\,t!}\bigg/\frac{(f+t)!}{f!\,t!} = \frac{f-t+1}{f+t} \qquad (t \leqslant f) \\ &= 0 \qquad (t > f). \end{aligned}$$

b) In order to be able to give immediate change to every customer it is necessary and sufficient that for any point in the line the number of people in front of that point (including the person at that point) who have only ten-dollar bills does not exceed by more than n the number of people who have fives. In other words, if $A_i = (x, y)$ we must have $y \leqslant x + n$. This means that paths corresponding to favorable outcomes lie entirely below or on the line $L_n : y - x = n$. Therefore the paths corresponding to unfavorable outcomes have a vertex on the line $L_{n+1} : y - x = n + 1$. If $t > f + n$, all the paths will have vertices on L_{n+1} since the origin lies below this line and $A_{f+t} = (f, t)$ lies above this line. Hence for $t > f + n$ the required probability is zero. If $t \leqslant n$ all outcomes are favorable and the required probability is 1.

We now study the remaining situation $n + 1 \leqslant t \leqslant f + n$. As in (a) it can be shown that the number of paths corresponding to the unfavorable outcomes is equal to the number of paths joining the point $A^n_0 = (-n-1,$

$n + 1$) (obtained by reflecting the point A_0 by the line L_{n+1}) to A_{f+t}. Therefore the total number of unfavorable outcomes is $\binom{f+t}{t-n-1}$ and so the total number of favorable outcomes is $\binom{f+t}{f} - \binom{f+t}{t-n-1}$. The required probability is therefore

$$\binom{f+t}{f} - \binom{f+t}{t-n-1}\Big/\binom{f+t}{f} = 1 - \frac{t(t-1)\cdots(t-n)}{(f+1)(f+2)\cdots(f+n+1)}.$$

Exercises

1 a) The numbers 1, 2, 3, 4, 5 are marked on five cards. Three cards are drawn at random (without replacement) and the digits thus obtained are written from left to right in the order in which they were drawn. What is the probability that the resulting number is even?

b) Answer part (a) under the assumption that the sampling is with replacement.

2 a) A monkey selects keys on a typewriter marked with $0, 1, 2, \ldots, 9, +, =$ at random. What is the probability that he will type $2 + 2 = 4$ if he types 5 symbols? What is the probability that he will type a valid equation of the form $x + y = z$?

b) Suppose the monkey only chooses the symbols 2, 4 $=$, $+$ at random. What is the probability that he types $2 + 2 = 4$ if he again types 5 symbols?

3 A box contains 100 good ball-bearings and 20 defective ball-bearings. If 15 ball-bearings are used, what is the probability that none are defective?

4 If r balls are placed in r cells, find the probability that exactly one cell remains empty.

5 a) If r men stand in a row what is the probability there will be exactly k men between 2 specified men X and Y?

b) Show that the probability is $\frac{1}{r-1}$ if they stand in a circle (consider only the arc leading counterclockwise from X to Y).

6 (The problem of Chévalier de Mérè) Which is more probable: Throwing at least one ace with four dice or at least a double ace in 24 throws of a pair of dice?

7 Find a probability problem whose solution will yield the following identity

$$1 + \frac{m-n}{m-1} + \frac{(m-n)(m-n-1)}{(m-1)(m-2)} + \cdots + \frac{(m-n)(m-n-1)\cdots 2\cdot 1}{(m-1)\cdots(n+1)(n)}$$

$$= \frac{m}{n} \quad (0 < n < m).$$

8 A drawer contains n pairs of gloves. If $2r$ gloves are chosen at random ($2r \leqslant n$), what is the probability that there will be exactly k pairs ($k = 0, 1, 2, \ldots, n$) among them?

9 From a group of m girls and n boys ($m > n$), one person after another is selected at random. Find the probability that at some point the number of boys and girls which have been selected will be the same.

10 A city has a population of k people. A person tells a rumor to a second person, who in turn repeats it to a third person etc. At each stage the listener is chosen at random from the $k - 1$ available people.

a) What is the proability that the rumor will be told n times without returning to the originator?

b) What is the probability that the rumor will not be repeated to any person?

c) Repeat the problem under the condition that at each stage the rumor is told by one person to N randomly chosen people at the same time.

11 Four children play with the same toys. Five toys were broken and out of these three were broken by the youngest child. Was the youngest child careless or could the frequency of his breakages be attributed to chance?

12 A boxing tournament has 4 participants. The number a boxer draws determines his opponent in the first rung of the tournament ladder (see diagram below).

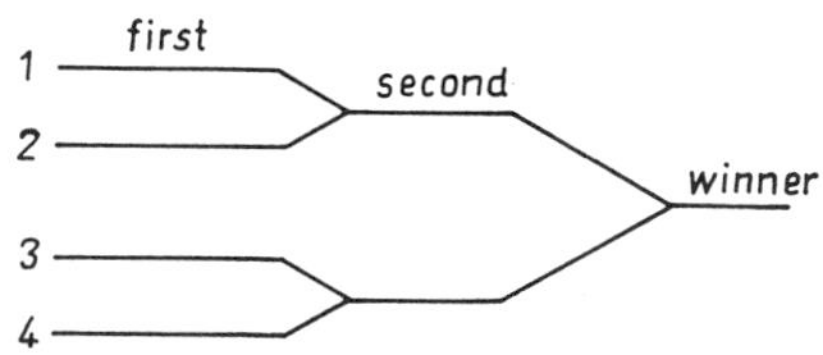

a) Suppose that the best boxer always defeats the next best and the latter always defeats all the rest. Find the probability that the second best boxer wins the second prize.

b) Rework (a) under the assumption that there are 2^n participants.

c) Suppose two brothers participate in situation (b). What is the probability that they meet in a match?

13 a) Let x, y, z, w be four nonnegative integers satisfying $x + y + z + w = 13$. What is the probability $P(x, y, z, w)$ that in a bridge game N, E, W, S have x, y, z, w diamonds respectively?

b) Find the probability that some player receives x, another y, and a third z and the fourth w diamonds if $x = y = z = 4$ $w = 1$, or $x = 5, y = 4, z = 3, w = 1$.

14 Twelve passengers randomly board a train consisting of 3 cars. What is the probability that:

a) there will be 4 people in the first car?

b) there will be 4 people in each car?

c) there will be 2 people in 1 car, 8 in another and 2 in the remaining car?

15 Solve problem 6 for $2n$ blades of grass.

16 A counterfeiter forms rolls of n coins by including m false coins. A suspicious bank clerk randomly draws 1 coin from each of the rolls. What is the probability of finding a false coin in the s rolls?

17 a) If k balls are selected at random from a box containing w white and b black balls what is the probability that exactly n of the drawn balls are white?

b) Use (a) to evaluate the sum

$$\binom{w}{0}\binom{b}{k} + \binom{w}{1}\binom{b}{k-1} + \binom{w}{2}\binom{b}{k-2} + \cdots + \binom{w}{k}\binom{b}{0}.$$

18 The twenty letters a_i and A_i ($i = 1, 2, \cdots, 10$) are marked on twenty cards. The cards are randomly arranged in a circle by first placing the 10 cards bearing the capital letters and then placing the cards bearing the small letters in the spaces between the first 10 cards. What is the probability that n small letters will be adjacent to the corresponding capital letter?

19 a) A train consists of n cars. Each of the r passengers selects at random the car in which he will sit. What is the probability that there will be at least 1 passenger in each carriage?

b) What is the probability that exactly k of the carriages will be occupied?

c) Use (a) to help evaluate the sum

$$\binom{n}{1} 1^r - \binom{n}{2} 2^r + \binom{n}{3} 3^r \cdots + (-1)^{n-1}\binom{n}{n} n^r.$$

20 a) Psychic powers of subjects are tested by asking the subject to guess the digits 0, 1, 2, 3, 4 which are marked on 5 cards of which they only see the backs. Find the probability of exactly i correct guesses.

b) The above experiment is repeated 5 times. What is the probability of more than 5 correct guesses out of 25 involved?

21 Show that for any integer m with $0 \leqslant m \leqslant n$ the probability, $P_{[m]}$, that exactly m among n events $A_1, A_2, \cdots, A_n$ occur simultaneously is given by

$$P_{[m]} = S_m - \binom{m+1}{m} S_{m+1} + \binom{m+2}{m} S_{m+2} - \cdots \pm \binom{n}{m} S_n,$$

where

$$S_0 = 1, \quad S_1 = \sum P(A_i),\ S_2 = \sum P(A_iA_j), \cdots, \quad S_n = P(A_1A_2 \cdots A_n).$$

22* Verify that the probability, P_m, that m or more of the events $A_1, \cdots, A_n$ occur simultaneously is given by

$$P_m = S_m - \binom{m}{m-1} S_{m+1} + \binom{m+1}{m-1} S_{m+2} - \binom{m+2}{m-1} S_{m+3} + \cdots + \binom{n-1}{m-1} S_n.$$

---

* The results in 21 and 22 remain valid even if the outcomes are not equally likely. (See [9])

CHAPTER VI

Möbius function and Polya's theorem

1 Partially ordered sets

NEARLY EVERYONE is familiar with the fact that the set $P = \{1, 2, 3, \ldots\}$ of natural numbers has an ordering relation $\geqslant$ with the following properties:

(1) $x \geqslant x$ for every x of P.

(2) If $x \geqslant y$ and $y \geqslant z$, then $x \geqslant z$.

(3) If $x \geqslant y$ and $y \geqslant x$, then $x = y$.

However there are many other sets of objects which have properties (1)–(3) provided that the relation $x \geqslant y$ is interpreted properly. To do this, we require the following concept.

DEFINITION 1 *A binary relation R on a set P is a statement $R(x, y)$ which is either true or false for every pair of elements x and y (not necessarily distinct) in P.*

Example 1

1 Let $P = \{f_1, m_1, d_{11}, d_{12}, s_{12}, f_2, m_2, s_{21}, s_{22}, d_{21}\}$, where f_i, m_i, d_{ij}, s_{ij} denote the father, mother, j-th daughter and j-th son of family i ($i = 1, 2$) respectively. Some binary relations on P are:

$R_1(x, y) \equiv x$ is the father of y.

$R_2(x, y) \equiv x$ is the mother of y.

$R_3(x, y) \equiv x$ is the brother of y.

$R_4(x, y) \equiv x$ is the sister of y.

The reader can easily verify that the condition of definition 1 is satisfied—for example, $R_1(d_{11}, f_1)$ is false, $R_1(f_1, d_{11})$ is true, $R_2(f_2, d_{11})$ is false, $R_3(s_{12}, d_{11})$ is true, $R_3(s_{12}, d_{21})$ is false and so on.

2 Let $P = \{1, 2, 3 \ldots\}$ be the set of natural numbers. Then $R_5(x, y) \equiv y$ divides x is a binary relation on P.

3 Let P be the set of all children in a school. Then

$$R_6(x, y) \equiv x \quad \text{is taller than } y$$

and

$$R_7(x, y) \equiv x \quad \text{is taller than or as tall as } y$$

are binary relations on P.

4 Let P be the set of objects represented by the points a_i in the diagram below.

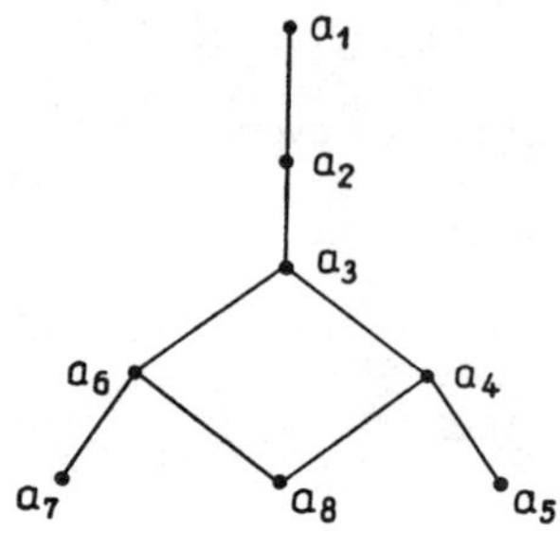

Figure 1

Let $R_8(x, y) \equiv$ either x and y are connected by a sequence of line segments and x lies above y in the diagram or x and y are identical. Then R_8 is a binary relation on P.

The notion of order can now be generalized as follows.

DEFINITION 2 *A partially ordered set P is a collection of elements $x, y, \ldots$ with an ordering relation R and an equality relation E such that axioms* (1)–(3) *hold when $x \geqslant y$ is replaced by $R(x, y)$ and $x = y$ is replaced by $E(x, y)$.*

A partially ordered set will sometimes be denoted by (P, R, E) for clarity.

The real numbers with the ordinary interpretation of $\geqslant$ and $=$ form a partially ordered set. Part 2 of example 1 provides us with an example of a partially ordered set (P, R_5, E) where $E(x, y)$ means $x = y$ in the usual sense. In part 3 of example 1, let $E(x, y) \equiv x$ is as tall as y; then (P, R_7, E) is a partially ordered set. However, R_6 cannot be used to form a partially ordered set since $R_6(x, x)$ is false. Finally, using the notation of part 4 of example 1, (P, R_8, E), where $E(x, y)$ means x is identical to y, is a partially ordered set.

None of the relations in part 1 of example 1 can be used to form a partially ordered set since $R_i(x, x)$ is false for $i = 1, 2, 3, 4$.

In the sequel, the relation $R(x, y)$ will be written as $x \geqslant y$ and $E(x, y)$ as $x = y$. We will also write $y \leqslant x$ as an alternate form of $x \geqslant y$, and $x > y$ (or $y < x$) if $x \geqslant y$ and $x \neq y$ ($E(x, y)$ is false).

The following is an important example of a partially ordered set which will be used later.

Example 2 Let S be a given finite set of objects. Form a set A from S by selecting some of the objects in S. Such a set is called a *subset* of S. We say that S *contains* A or alternatively that A is *contained* in S.

Let P be the set whose elements are all possible subsets of S. For example, if $S = \{a, b, c\}$ then $P = \{0, \{a\}, \{b\}, \{c\}, \{a, b\}, \{a, c\}, \{b, c\}, S\}$. Here 0 denotes the void subset of S; that is, the subset which contains no elements of S. The set S itself is frequently denoted by 1. $x \leqslant y$ means that x is a subset of y; for example, $0 \leqslant \{a\}$, $0 \leqslant \{b\}$, $0 \leqslant \{c\}$*, $\{a\} \leqslant \{a, b\}$, $\{a\} \leqslant \{a, c\}$, $\{b\} \leqslant \{b, c\}$, $\{b\} \leqslant \{a, b\}$, $\{c\} \leqslant \{a, c\}$ and $\{a, b\} \leqslant \{a, b, c\} = 1$. Equality of two sets means that the sets are identical, that is, they are composed of exactly the same objects; the order in which the objects appear is immaterial. For example, $\{a, b\} = \{b, a\}$, $\{a, b, c\} = \{c, b, a\}$ while $\{a\} \neq \{b, c\}$ and $\{a, b\} \neq \{a, b, c\}$. $(P, \leqslant, =)$ is a partially ordered set and will be said to be ordered by inclusion. $x \leqslant y$ is read as x is contained in y or x is a subset of y.

Note that in a partially ordered set two elements need not be comparable, that is, both $x \leqslant y$ and $y \leqslant x$ may not hold. In the partially ordered system constructed from part 4 of example 1, a_8 and a_7 are not comparable.

DEFINITION 3 *A partially ordered set which also satisfies:*
(4) *If x, y are elements of P, then either $x \leqslant y$ or $y \leqslant x$ is called a* chain (*or* simply ordered set).

The natural numbers with their natural order is a chain. The set of elements $\{a_1, a_2, a_3, a_6, a_8\}$ with the order described by figure 1 is a chain.

Much of the terminology which is used for the real number system with the greater than or equal to relation is carried over to partially ordered systems.

DEFINITION 4 *If S is a subset of a partially ordered set P, then an element x of P such that $x \geqslant s$ for every s of S is called an* upper bound *of S. If z is an upper bound of S such that $z \leqslant x$ for every upper bound x of S, then z is called the* least upper bound *of S.*

---

* By convention, the empty set is regarded as a subset of every set.

The subset $S = \{a_3, a_4, a_5\}$ of the partially ordered system described by figure 1 has upper bounds a_1, a_2 and a_3; its least upper bound is a_3. Let P be the natural numbers with their natural order and S be the subset of all odd integers. Then S does not have an upper bound.

If a set has an upper bound it does not necessarily have a least upper bound. For example, let P be the set of all rational numbers and S be all rational numbers less than $\sqrt{2}$. Under the usual ordering relation, S has an upper bound but does not have a least upper bound.

LEMMA 1 *If S has a least upper bound then it is unique.*

Proof Let x and y be two least upper bounds of S. Then, by definition, $x \leqslant y$ and $y \leqslant x$. Hence $x = y$ by axiom (3).

DEFINITION 5 *If S is a subset of a partially ordered set P, then an element x of P such that $x \leqslant s$ for every s of S is called a* lower bound *of S. If z is a lower bound of S, such that $x \leqslant z$ for every lower bound x of S, then z is called a* greatest lower bound *of S.*

It follows that if S has a greatest lower bound, it is necessarily unique.

DEFINITION 6 *The greatest lower bound of P is called the* zero element *and its least upper bound is called the* unit element (*or* all element).

The zero and unit elements need not exist. The partially ordered set formed from the natural numbers using the natural order has no all element while its zero element is 1. The system described in figure 1 has a_1 as its unit element but it has no zero element.

DEFINITION 7 *An* interval $[x, y]$ *where $x \leqslant y$ is the set of all elements z such that $x \leqslant z \leqslant y$.*

The interval [5, 8] of the natural numbers is the set $\{5, 6, 7, 8\}$ while the interval $[a_3, a_8]$ in figure 1 is the set $\{a_3, a_6, a_4, a_8\}$.

DEFINITION 8 *A partially ordered set P is said to be* locally finite *if the number of elements in every interval $[x, y]$ is finite.*

A partially ordered set with a finite number of elements is obviously locally finite. The natural numbers with the usual ordering is locally finite while the real numbers are not locally finite.

Exercises

1 Draw diagrams similar to figure 1 for each of the following partially ordered systems:
a) the set of all subsets of a given set of four points.

b) the numbers 1, 2, 3, 4, 6, 8, 12, 24 under their natural order.
c) the numbers in (b) under divisibility.

2 Which of the following sets are partially ordered systems?
a) All pairs of numbers (x, y), if $(x, y) \leqq (x', y')$ means that $x' \leqslant x$ and $y \leqslant y'$.
b) All pairs of numbers (x, y) if $(x, y) \leqslant (x', y')$ means that $x + x' \leqslant x$ and $y \leqslant y'$.
c) All pairs of numbers (x, y) if $(x, y) \leqslant (x', y')$ means that either $x < x'$ or $x = x'$ and $y \leqslant y'$.
d) All subsets of a finite set S if $A \leqslant B$ means that B is contained in A (i.e. B is a subset of A).

3 Show that in the partially ordered system of the natural numbers under division, the least common multiple of a finite subset S is the least upper bound of S while the greatest common divisor is the greatest lower bound of S.

2 The Möbius function

Consider the class $\mathscr{C}$ of all real valued functions f defined for x, y in P, a locally finite partially ordered set, such that $f(x, y) = 0$ if $x > y$ or if x and y are not comparable. The following functions belong to $\mathscr{C}$.

DEFINITION 1 *The* Kronecker delta function, δ, *is defined by* $\delta(x, x) = 1$ *and* $\delta(x, y) = 0$ *if* $x \neq y$. *The* zeta function, ζ, *is defined by* $\zeta(x, y) = 1$ *for* $x \leqslant y$, $\zeta(x, y) = 0$ *otherwise.*

Example 1 For the partially ordered set depicted by the following diagram

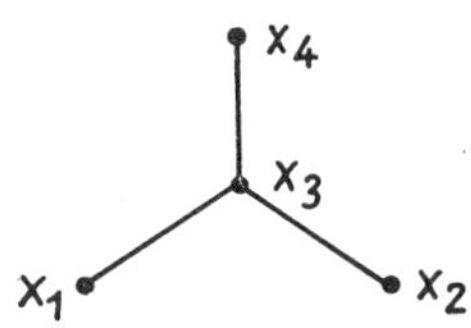

$$1 = \zeta(x_1, x_1) = \zeta(x_2, x_2) = \zeta(x_3, x_3) = \zeta(x_4, x_4) = \zeta(x_1, x_3) = \zeta(x_1, x_4)$$
$$= \zeta(x_2, x_3) = \zeta(x_2, x_4) = \zeta(x_3, x_4) \text{ and 0 otherwise while } \delta(x_i, x_i)$$
$$= 1 \ (i = 1, 2, 3, 4) \text{ and 0 otherwise.}$$

Later we shall see that given f in $\mathscr{C}$, it is important to be able to find functions g and h in $\mathscr{C}$ such that for fixed x and y,

$$\delta(x, y) = \sum_{x \leq z \leq y} f(x, z)\, g(z, y) \tag{1}$$

$$\delta(x, y) = \sum_{x \leq z \leq y} h(x, z) f(z, y). \tag{2}$$

Since P is locally finite, the sum on the right is finite.

LEMMA 1 *Given f, functions g and h in* $\mathscr{C}$ *satisfying* (1) *and* (2) *can be found if and only if for every x in P,* $f(x, x) \neq 0$.

Proof We shall prove the result for (1); the proof for (2) is similar. Suppose that there exists a solution g in $\mathscr{C}$ of (1). This requires that $1 = \delta(x, x) = f(x, x)\, g(x, x)$ for every x. Hence $f(x, x) \neq 0$ for every x in P is necessary.

Next suppose that for every x in P, $f(x, x) \neq 0$. Then $g(x, x) = f(x, x)^{-1}$ for every x. Since P is locally finite, there is only a finite number of elements in the interval $[x, y]$. Suppose $x \neq y$. Then there exists an element w_1 which is a predecessor of y; that is, $w_1 < y$ and there is no element z in P such that $w_1 < z < y$. For such an element w_1, we can find $g(w_1, y)$. Since

$$\delta(w_1, y) = 0 = \sum_{w_1 \leq z \leq y} f(w_1, z)\, g(z, y)$$

it follows that

$$-f(w_1, w_1)\, g(w_1, y) = f(w_1, y)\, g(y, y)$$

and we can solve for $g(w_1, y)$ since $f(w_1, w_1) \neq 0$ and $g(y, y)$ is known.

Let $\{w_{1i}\}$, $i = 1, \dots, n$, be the set of all predecessors of y and let w_2 be a predecessor of w_{1j} for $j = 1, 2, \dots, m$. Then, as described above, $g(w_2, w_{1j})$ can be found. Since

$$0 = \sum_{w_2 \leq z \leq y} f(w_2, z)\, g(z, y)$$

it can easily be seen that

$$-f(w_2, w_2)\, g(w_2, y) = \sum_{j=1}^{m} f(w_2, w_{1j})\, g(w_{1j}, y) + f(w_2, y)\, g(y, y).$$

This last equation determines $g(w_2, y)$ since every term on the right hand side is known and $f(w_2, w_2) \neq 0$. Repeating this procedure*, we can eventually calculate $g(x, y)$ because $[x, y]$ is locally finite. We set $g(x, y) = 0$ if $x \leqslant y$ is not valid; then g is in C.

LEMMA 2 *If g is a solution of* (1) *and h is a solution of* (2), *then* $g = h$.

Proof From the definition of δ *it follows that*

$$h(x, y) = \sum_{x \leqslant w \leqslant y} h(x, w)\, \delta(w, y). \tag{3}$$

---

* This result can be proved rigorously by induction.

Replacing $\delta(w, y)$ by its value given by (1) in (3), (3) reduces to

$$h(x, y) = \sum_{x \leqslant w \leqslant y} h(x, w) \sum_{w \leqslant z \leqslant y} f(w, z)\, g(z, y)$$

$$= \sum_{x \leqslant w \leqslant y} h(x, w) \sum_{x \leqslant z \leqslant y} f(w, z)\, g(z, y)$$

$$\text{(since } f(w, z) = 0 \text{ for } x \leqslant z < w)$$

$$= \sum_{x \leqslant z \leqslant y} g(z, y) \sum_{x \leqslant w \leqslant y} h(x, w) f(w, z)$$

$$= \sum_{x \leqslant z \leqslant y} g(z, y) \sum_{x \leqslant w \leqslant z} h(x, w) f(w, z)$$

$$\text{(since } f(w, z) = 0 \text{ for } z < w \leqslant y)$$

$$= \sum_{x \leqslant z \leqslant y} g(z, y)\, \delta(x, z) = g(x, y) \quad \text{(from (2))}.$$

Summarizing, we see that lemmas (1) and (2) assert that given any f in $\mathscr{C}$ such that $f(x, x) \neq 0$ for all x, there exists a unique function g in $\mathscr{C}$ satisfying (1) and (2)*. Since $\zeta(x, x) = 1 \neq 0$, the following definition is sensible.

DEFINITION 2 *The Möbius function* $\mu(x, y)$ *is the unique solution of* (1) *and* (2) *for* $f(x, y) = \zeta(x, y)$, *the zeta function.*

Hence

$$\mu(x, x) = 1, \quad \text{for every } x \text{ of } P; \tag{4}$$

$$\mu(x, y) = -\sum_{x \leqslant z < y} \mu(x, z), \quad x < y \text{ fixed}; \tag{5}$$

$$\mu(x, y) = -\sum_{x < z \leqslant y} \mu(z, y), \quad x < y \text{ fixed}; \tag{6}$$

where (5) follows from (2) while (6) follows from (1).

These last equations are used to calculate the values of the Möbius function.

Example 2

1 The partially ordered set with two distinct elements x, y satisfying $x \leqslant y$ has the following Möbius function:

$$\mu(x, x) = \mu(y, y) = 1; \quad \mu(x, y) = -1.$$

The last equation is an immediate consequence of (6).

* When the sum of two functions in $\mathscr{C}$ and the multiplication of a function by scalars is defined in the usual manner and when the product $h = fg$ is defined by the right side of (1), the functions of $\mathscr{C}$ form the *incidence algebra* $A(P)$ of P. The delta function is the identity element. Lemma 1 states that if $f(x, x) \neq 0$, it has a right and left inverse. Lemma 2 is just the well known fact that in such situations there exists a unique inverse. The Möbius function is just the inverse of the zeta function.

2 We now calculate the Möbius function of the partially ordered set in example 1. From (4), $\mu(x_i, x_i) = 1$ for $i = 1, 2, 3, 4$ and from (6)

$$\mu(x_3, x_4) = -\sum_{x_3 < z \leqslant x_4} \mu(z, x_4) = -\mu(x_4, x_4) = -1.$$

Similarly,

$$\mu(x_2, x_3) = \mu(x_1, x_3) = -1.$$

Another application of (6) yields

$$\mu(x_2, x_4) = -\sum_{x_2 < z \leqslant x_4} \mu(z, x_4) = -\mu(x_3, x_4) = 0$$

and $\mu(x_1, x_4) = 0$.

3 Let P be the partially ordered set of the positive integers, where $x \leqslant y$ means that x divides y (written $x|y$). If $x \leqslant z \leqslant y$, then $z = xd$ and $y = zk$ where k and d are positive integers. Hence $y/x = dk$ which shows that d is a divisor of y/x. Therefore, the defining relation for the Möbius function is

$$\delta(x, y) = \sum_{x \leqslant z \leqslant y} \mu(x, z) = \sum_{d \left| \frac{y}{x} \right.} \mu(x, xd).$$

Setting $y/x = n$, the last equation reduces to

$$\sum_{d|n} \mu(x, xd) = \begin{cases} 1, & \text{if } n = 1, \\ 0, & \text{if } n > 1; \end{cases} \tag{7}$$

the sum is over all positive divisors d of n.

By lemmas 1 and 2 this relation determines the Möbius function uniquely. Hence if we can find any function of d satisfying (7) it must be the Möbius function.

There is a well known function, studied in the theory of numbers, which satisfies (7). This function, defined for positive integers n, is also called the Möbius function and is denoted $\mu(n)$. Every positive integer $n > 1$ has a unique factorization as a product of prime powers

$$n = p_1^{i_1} p_2^{i_2} \cdots p_r^{i_r} \tag{8}$$

where the p's are different primes and the i's are positive integers. $\mu(n)$ is defined by

$$\mu(n) = \begin{cases} 1, & \text{if } n = 1, \\ 0, & \text{if any } i_k > 1 \text{ in (8)}, \\ (-1)^r, & \text{if } i_1 = i_2 = = \cdots = i_r = 1 \text{ in (8)}. \end{cases} \tag{9}$$

LEMMA 3

$$\sum_{d|n} \mu(d) = \begin{cases} 1, & \text{if } n = 1, \\ 0, & \text{if } n > 1. \end{cases}$$

Proof If $n = 1$, then $d = 1$ is the only divisor, and $\mu(1) = 1$. Suppose that $n > 1$ and is given by (8). Let $m = p_1 p_2 \cdots p_r$. A divisoı d of n that is not a divisor of m must have a repeated prime factor and so $\mu(d) = 0$. Therefore, $\sum_{d|n} \mu(d) = \sum_{d|m} \mu(d)$. To evaluate this last sum, note that there are $\binom{r}{k}$ divisors of m which are the product of k distinct primes and for each of these, $\mu(d) = (-1)^k$. Hence,

$$\sum_{d|m} \mu(d) = 1 - \binom{r}{1} + \binom{r}{2} + \cdots + (-1)^r \binom{r}{r} = (1 - 1)^r = 0.$$

Lemma 3 and (7) imply that $\mu(x, y) = \mu(y/x)$.

4 Let P be the partially ordered set of all subsets of a finite set S ordered by inclusion (see example 1.2). We will show that if $x \leqslant y$, then

$$\mu(x, y) = (-1)^{n(y)-n(x)} \tag{10}$$

where $n(x)$ and $n(y)$ are the number of elements of S in x and y respectively.

The result is valid when $n(y) - n(x) = 0$ or 1. If $n(y) - n(x) = 2$, there are three subsets z such that $x \leqslant z < y$. Hence from (5),

$$\mu(x, y) = -(1 - 2) = (-1)^2.$$

Next, suppose that $n(y) - n(x) = 3$. There are $\binom{3}{j}$ subsets z such that $x \leqslant z < y$ with $n(z) - n(x) = j$. These are obtained by adjoining to x, j of the 3 elements of y not in x. Once again, (5) yields

$$\mu(x, y) = -\left[1 + \binom{3}{1}(-1) + \binom{3}{2}(-1)^2\right]$$

$$= -[(1 - 1)^3 - (-1)^3] = (-1)^3.$$

as can easily be seen by expanding $(1 - 1)^3$ by the binomial theorem. Since $\mu(x, y)$ is known for $n(y) - n(x) = i$ $(i = 1, 2, 3)$ we can calculate $\mu(x, y)$

for $n(y) - n(x) = 4$ from (5). Repeating this procedure*; it can be shown that (10) is valid for any integral value of $n(y) - n(x)$.

The next result is known as the *Möbius inversion theorem.*

THEOREM 1 *Let P be a locally finite partially ordered set with a zero element and $f(x)$ a function defined on P. Define $g(x)$ on P by*

$$g(x) = \sum_{y \leqslant x} f(y). \tag{11}$$

Then, if μ is the Möbius function of P,

$$f(x) = \sum_{y \leqslant x} g(y)\,\mu(y, x) \tag{12}$$

for all x in P. Conversely, (12) *implies* (11).

Proof The sums in (11) and (12) are well defined since there is only a finite number of elements in $[0, x]$. Substituting from (11) in the right side of (12) yields

$$\sum_{y \leqslant x} g(y)\,\mu(y, x) = \sum_{y \leqslant x}\sum_{z \leqslant y} f(z)\,\mu(y, x) = \sum_{y \leqslant x}\Big(\sum_{z} f(z)\,\zeta(z, y)\Big)\mu(y, x).$$

Interchanging the order of summation, this becomes

$$\begin{aligned}\sum_z f(z) \sum_{y \leqslant x} \zeta(z, y)\,\mu(y, x) &= \sum_z f(z) \sum_{z<y<x} \zeta(z, y)\,\mu(y, x)\\ &= \sum_z f(z)\,\delta(z, x) = f(x)\end{aligned}$$

which proves the validity of (12) under the hypothesis that (11) is valid.

Conversely, assume that (12) holds. Then

$$\begin{aligned}\sum_{y \leqslant x} f(y) = \sum_{y \leqslant x}\sum_{z \leqslant y} g(z)\,\mu(z, y) &= \sum_z g(z) \sum_{z \leqslant y \leqslant x} \mu(z, y)\\ = \sum_z g(z)\,\delta(z, x) &= g(x)\end{aligned}$$

since $\mu(z, y) = 0$ for $z \leqslant y$.

As an application of the Möbius inversion theorem, we deduce the following result.

THEOREM 2 *Let f and g be functions defined for every positive integer satisfying*

$$g(n) = \sum_{d \mid n} f(d). \tag{13}$$

---

* A rigorous proof requires induction.

Then

$$f(n) = \sum_{d|n} \mu(d)\, g\left(\frac{n}{d}\right). \tag{14}$$

Conversely, (14) *implies* (13).

Proof Using the notation of part 3 of example 2 with $x = n$ and $y = d$, (13) is an immediate consequence of (11), while (12) reduces to

$$f(n) = \sum_{d|n} g(d)\,\mu\left(\frac{n}{d}\right) = \sum_{d|n} g\left(\frac{n}{d}\right)\mu(d),$$

which is just (14).

This last theorem can be used to enumerate circular permutations.

Example 3 By a *circular permutation,* we mean an arrangement of the n letters $a_1, a_2, \ldots, a_n$ in a circle (the letters need not be distinct). Two circular permutations are considered to be the same if by shifting the letters of one of the arrangements clockwise around the circle (without changing their order) the second arrangement is obtained (see example 1.3.11).

In order to find the number of circular permutations of n letters which can be formed from k distinct letters, associate with each circular permutation a number of linear sequences in the following manner: write the circular arrangement in a row beginning with a fixed letter and copy the remaining letters in the order in which they appear when the circle is traversed clockwise. For example, if the letters $a_1, a_2, \ldots, a_n$ are arranged in a circle with a_1 following a_n the following linear sequences are obtained:

$$a_1, a_2, a_3, \ldots, a_n; \quad a_2, a_3, \ldots, a_n, a_1; \quad a_3, \ldots, a_n, a_1, a_2; \quad \ldots;$$
$$a_n, a_1, \ldots, a_{n-1}$$

Further examples appear in the diagram below.

$n = 6 \quad d = 3$	$n = 6 \quad d = 2$
a	a
$c \quad\quad b$	$b \quad\quad b$
$b \quad\quad c$	$a \quad\quad a$
a	b
abcabc	*ababab*
bcabca	*bababa*
cabcab	

Figure 1

Conversely, the linear sequences in each of the above sets all determine the same circular permutation. However, not all n linear sequences corresponding to the same circular permutation need be different. If the sequence $a_1, a_2, \ldots, a_n$ consists of a sequence of d distinct letters $a_1, a_2, \ldots, a_d$ repeated n/d times, the linear sequences repeat after the first d letters (see figure 1 for $n = 6$, $d = 3$ and $n = 6$, $d = 2$). Such a circular permutation will be said to be of *period d*.

Using k different letters, k^n distinct linear sequences of length n can be formed. We now calculate the total number of linear sequences in another manner. Let $M(d)$, where $d|n$, be the number of different circular permutations of period d. Then $dM(d)$ is the total number of distinct linear sequences corresponding to circular permutations of period d. Since each linear sequence is uniquely associated with a circular permutation of period d (for some d),

$$\sum_{d|n} dM(d) = k^n. \tag{15}$$

Letting $g(x) = k^x$ and $f(x) = xM(x)$, we may apply (14) to (15) and obtain

$$nM(n) = \sum_{d|n} \mu(d)\, k^{n/d}$$

or

$$M(n) = \frac{1}{n} \sum_{d|n} \mu(d)\, k^{n/d}.$$

This gives the number of circular permutations of period n. The total number, $T(n)$, of circular permutations of length n is

$$T(n) = \sum_{d|n} M(d).$$

As another application of Möbius inversion, we will derive theorem 4.2.1.

Example 4 Let I be the set of integers $1, 2, \ldots, n$ and let the properties $A_1, A_2, \ldots, A_n$ be associated with these integers. Let S be a finite set of objects each of which has some of the properties A_i and let x be a fixed subset of I containing $n - m$ of the integers in I. Let $f(x)$ denote the number of elements in S having exactly m ($0 \leqslant m \leqslant n$) of the properties A_i for i not in x. Using the partial ordering described in example 1.2, the function

$$g(x) = \sum_{y \leqslant x} f(y)$$

is the number of elements in S having all the properties A_i for i not in x and possibly others. The inversion formula (12) yields

$$f(x) = \sum_{y \leqslant x} (-1)^{n(x)-n(y)} g(y) \tag{16}$$

upon substituting from (10) for the Möbius function. Let the integers which are in x be denoted by $i_1, i_2, \ldots, i_{n-m}$ and those which are not in x by $j_1, j_2, \ldots, j_m$. Then, using the notation of theorem 4.2.1, (16) can be rewritten as

$$\begin{aligned} f(x) = N(A_{j_1}, A_{j_2}, \ldots, A_{j_m}) + \cdots + (-1)^r \times \\ \times \sum_{(k_1, k_2, \ldots, k_r)} N(A_{j_1}, \ldots, A_{j_m} A_{k_1}, \ldots, A_{k_r}) + \cdots \\ + (-1)^n N(A_{j_1}, \ldots, A_{j_m}, A_{i_1}, \ldots, A_{i_{n-m}}) \end{aligned} \tag{17}$$

where $\sum_{(k_1, \ldots, k_r)}$ denotes summation over all r-combinations $(k_1, \ldots, k_r)$ which can be formed from the integers $i_1, i_2, \ldots, i_{n-m}$. Summing (17) over all possible subsets x of I containing $n - m$ integers yields theorem 4.2.1 since $N_{[m]} = \sum_x f(x)$, $\sum_x N(A_{j_1}, \ldots, A_{j_m}) = S_m$ and

$$\sum_x \sum_{(k_1, \ldots, k_r)} N(A_{j_1}, A_{j_2}, \ldots, A_{j_m}, A_{k_1}, A_{k_2}, \ldots, A_{k_r}) = \binom{m+r}{m} S_{m+r}.$$

The last assertion follows from the fact that every term in S_{m+r} will be repeated $\binom{m+r}{m}$ times, since a term such as $N(A_{j_1}, A_{j_2}, \ldots, A_{j_m}, A_{k_1}, \ldots, A_{k_r})$ arises from any subset x which does not contain any m of the $m + r$ integers $j_1, j_2, \ldots, j_m, k_1, \ldots, k_r$.

Exercises

1 a) Find the Möbius function for the set of positive integers with their natural order.
b) Find the Möbius functions of the following partially ordered sets:

2 a) Prove theorem 1 replacing the assumption that P has a zero element by the assumption that there exists an element p of P such that $f(x) = 0$ unless $x \geqslant p$.

b) Let f be a function defined on a locally finite partially ordered set P. Suppose there is an element q such that $r(x)$ vanishes unless $x \leqslant q$. Show that if $g(x) = \sum_{y \geqslant x} f(y)$ then $f(x) = \sum_{y \geqslant x} \mu(x, y)\, g(y)$.

3 Let x and y be two elements of a locally finite partially ordered set P. Define a partially ordered set P^* by setting $x \leqslant {}^*y$ if and only if $x \leqslant y$. Let μ^* denote the Möbius function of P^*. Show that $\mu^*(x, y) = \mu(y, x)$.

4 The function f is defined on the positive integers n by $\log n = \sum_{d|n} f(d)$. Prove that $f(n) = \log p$ if $n = p^i$, where p is a prime and i is a positive integer, and $f(n) = 0$ otherwise.

5 A Dirichlet generating function $A(x)$ for a sequence of numbers $\{a_i\}$ is the formal series $A(x) = \sum_{n=1}^{\infty} a_n/n^x$. If $B(x) = \sum_{n=1}^{\infty} \frac{b_n}{n^x}$, we define

$$A(x) + B(x) = C(x) = \sum_{n=1}^{\infty} \frac{c_n}{n^x} \qquad (c_n = a_n + b_n)$$

and

$$A(x)\, B(x) = D(x) = \sum_{n=1}^{\infty} \frac{d_n}{n^x} \qquad \left(d_n = \sum_{d|n} \frac{a_d b_n}{d}\right).$$

a) Prove that $AB = BA$ and $A(BC) = (AB)C$ for any three generating functions A, B and C.

b) Let $I(x) = 1/n^x$. Show that $AI = IA = A$.

c) Show that if $a_1 \neq 0$, the series $A(x)$ has an inverse $B(x)$ satisfying $A(x)\, B(x) = I(x)$.

d) The Riemann zeta function is defined by $\zeta(x) = \sum_{n=1}^{\infty} \frac{1}{n^x}$. Show that the inverse of $\zeta(x)$ is the series $\zeta(x)^{-1} = \sum_{n=1}^{\infty} \mu(n)/n^x$, where $\mu(n)$ is the Möbius function defined in (9). Let $a_n = f(n)$ and $b_n = g(n)$ for all n and $B(x) = A(x)\,\zeta(x)$; then $A(x) = B(x)\,\zeta(x)^{-1}$. How is this related to the Möbius inversion formula?

3 Groups

The notation of a group requires the introduction of the following concept.

DEFINITION 1 *A* (binary) operation $\circ$ *on a set S is a rule which associates with any two elements a and b in S a unique element c. This is denoted symbolically by $a \circ b = c$. The element c does not necessarily belong to S. If, for* *arbitrary elements a and b in S the element c is also in S, then S is said to be* closed (*under the operation* $\circ$).

Example 1

1 Let S be the set of nonnegative integers.

a) if $\circ$ is the ordinary operation of addition, then S is closed.

b) Let $\circ$ denote the operation of subtraction. Then S is not closed since $2 \circ 3 = -1$ is not in S.

c) If $\circ$ denotes the operation of division, then S is not closed, since $2/4$ is not in S.

2 Let S be the set of all integers; S is closed under addition, multiplication and subtraction. However, S is not closed under division, since $2/3$ is not an integer.

3 Let S be the set of real numbers. Then S is closed under addition, subtraction and multiplication but not under division since division by zero is meaningless. The set $\bar{S}$ of all real numbers except zero is closed under division.

There are many binary operations other than the ordinary operations of arithmetic.

Example 2 A rotation in a plane of β degrees of a unit radius vector about an origin 0 measured counterclockwise from a given initial position of the radius vector will be denoted by R_β (see figure below).

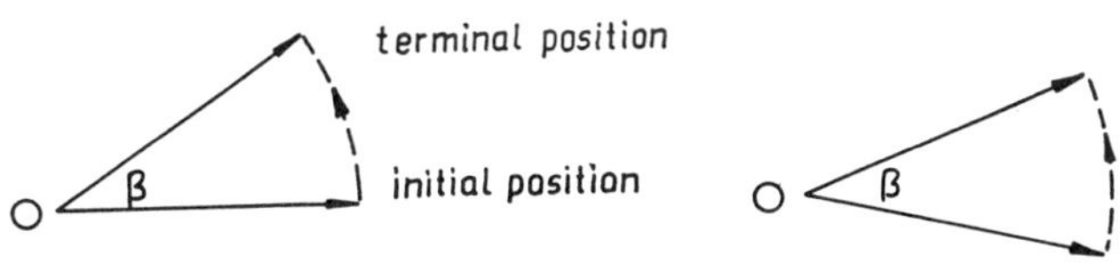

Figure 1

Let S be the set of all such rotations. $R_\alpha \circ R_\beta$ is defined to be the rotation produced by first rotating through an angle β and then rotating through an angle α. S is closed under $\circ$, in fact, $R_\alpha \circ R_\beta = R_{\alpha+\beta}$.

Example 3 A binary operation can be defined on the set S of 8 motions which carry the square into itself. Imagine a cardboard square laid on a plane with fixed axes so that the center of the square coincides with the origin and one side is horizontal. The square is carried into itself by the following motions:

1 R: a 90° rotation counterclockwise around its center 0.
2 R_1: a 180° rotation counterclockwise around 0.
3 R_2: a 270° rotation counterclockwise around 0.
4 H: a reflection in the horizontal axis through 0.
5 V: a reflection in the vertical axis through 0.
6 D_1: a reflection in the diagonal which lies in quadrants I and III.
7 D_2: a reflection in the diagonal which lies in quadrants II and IV.
8 I: the motion which leaves every point fixed.

These motions can be represented by diagrams of the following type:

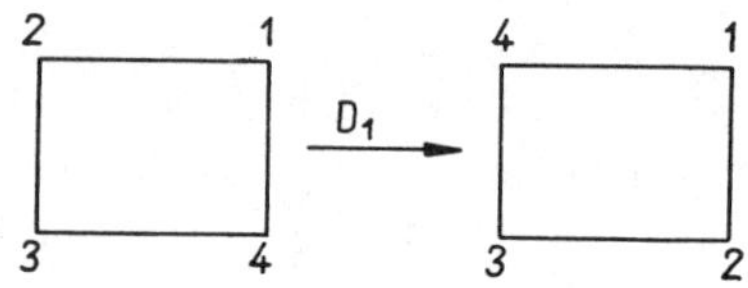

Figure 2

The square on the left is transformed by D_1 into the square on the right. Vertices 1 and 3 are fixed while 2 and 4 are interchanged.

Let M_1 and M_2 be any two motions; $M_1 \circ M_2$ is defined to be the performance of M_2 followed by the performance of M_1. Thus $R \circ H$ is obtained by first reflecting the square in a horizontal axis, then rotating counterclockwise through 90° (see figure 3)

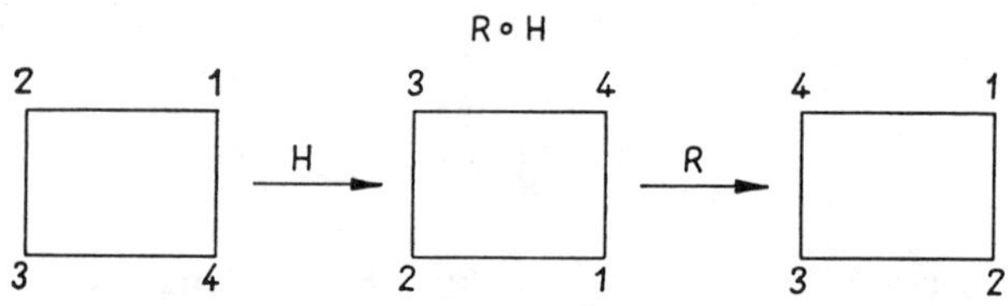

Figure 3

From figures 2 and 3 it follows that $R \circ H = D_1$. Similarly, $D_1 \circ D_2$ is defined as the result of a reflection in the diagonal which lies in quadrants II and IV, followed by a reflection in the diagonal which lies in quadrants I and III. Hence $D_1 \circ D_2 = R_1$. In the table below, the entry in the intersection of the row labelled M_i and the column labelled M_j is the motion $M_i \circ M_j$.

	R	R_1	R_2	H	V	D_1	D_2	I
R	R_1	R_2	I	D_1	D_2	V	H	R
R_1	R_2	I	R	V	H	D_2	D_1	R_1
R_2	I	R	R_1	D_2	D_1	H	V	R_2
H	D_2	V	D_1	I	R_1	R_2	R	H
V	D_1	H	D_2	R_1	I	R	R_2	V
D_1	H	D_2	V	R	R_2	I	R_1	D_1
D_2	V	D_1	H	R_2	R	R_1	I	D_2
I	R	R_1	R_2	H	V	D_1	D_2	I

The above table is called a *multiplication table* for the group.

Example 4 Recall that a *permutation* on n symbols $1, 2, 3, \ldots, n$ is a rearrangement of the said symbols whereby any symbol i is replaced by a symbol i' (the i' are all distinct). A permutation π is usually represented by a 2-line symbol like

$$\pi = \begin{pmatrix} 1 & 2 & 3 & \cdots & n \\ 1' & 2' & 3' & \cdots & n' \end{pmatrix}$$

where the lower line gives the symbols replacing those in the upper line. In other words, π transforms $(1, 2, 3, \ldots, n)$ into $(1', 2', 3', \ldots, n')$. A binary operation can be defined on the set of all $n!$ permutations by defining $\pi_1 \circ \pi_2 \equiv \pi_1\pi_2$ as the permutation resulting from their successive application. Thus, for $n = 4$,

$$\begin{pmatrix} 1 & 2 & 3 & 4 \\ 4 & 1 & 3 & 2 \end{pmatrix}\begin{pmatrix} 1 & 2 & 3 & 4 \\ 1 & 3 & 4 & 2 \end{pmatrix} = \begin{pmatrix} 1 & 2 & 3 & 4 \\ 4 & 3 & 2 & 1 \end{pmatrix}$$

since $1 \to 1 \to 4$, $2 \to 3 \to 3$, $3 \to 4 \to 2$ and $4 \to 2 \to 1$. On the other hand,

$$\begin{pmatrix} 1 & 2 & 3 & 4 \\ 1 & 3 & 4 & 2 \end{pmatrix}\begin{pmatrix} 1 & 2 & 3 & 4 \\ 4 & 1 & 3 & 2 \end{pmatrix} = \begin{pmatrix} 1 & 2 & 3 & 4 \\ 2 & 1 & 4 & 3 \end{pmatrix}.$$

Some binary operations have special properties. In definitions 2–5, a, b, c, d and e are elements of a set S on which there is a binary operation $\circ$.

DEFINITION 2 *The operation* $\circ$ *is said to be* commutative *if* $a \circ b = b \circ a$ *for any elements* a *and* b *in* S.

In part 3 of example 1, the operations of addition and multiplication are commutative while the operations of subtraction and division are not commutative. The operations defined in examples 3 and 4 aie not commutative. The operations defined in examples 3 and 4 are not commutative while the operation defined in example 2 is commutative.

DEFINITION 3 *The operation* $\circ$ *is* associative *if*

$$(a \circ b) \circ c = a \circ (b \circ c),$$

for any elements a, b *and* c *of a closed set* S.

In part 3 of example 1, the operation of addition is associative—for example, $(2 \circ 3) \circ 4 = 5 \circ 4 = 9$ while $2 \circ (3 \circ 4) = 2 \circ 7 = 9$. The operation of multiplication is also associative—for instance, $(2 \circ 3) \circ 4 = 6 \circ 4 = 24$ and $2 \circ (3 \circ 4) = 2 \circ 12 = 24$. The operation of subtraction is not associative since $(2 \circ 3) \circ 4 = -1 \circ 4 = -5$ while $2 \circ (3 \circ 4) = 2 \circ (-1) = 3$. Division is also not associative since $(2 \circ 3) \circ 4 = (2/3) \circ 4 = 1/6$ while $2 \circ (3 \circ 4) = 2 \circ (3/4) = 8/3$. The associativity of the operation appearing in example 2 follows from the associativity of addition.

The associativity of the operations in examples 3 and 4 can be proved by straightforward but tedious computation. It follows easily from the more general result derived in the next example.

Example 5 A *mapping** $\Phi : S \to T$ from a nonempty set S *into* a set T is a rule which assigns to each element s in S a unique image element $\Phi(s)$ in T (see figure 4).

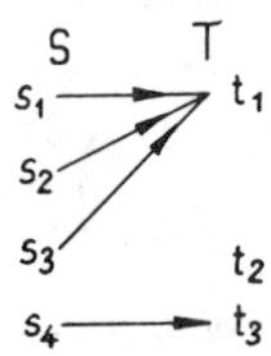

Figure 4

* Also called transformation, function, and operator.

Notice that it is possible for 2 different elements s_1 and s_2 to have the same image in T, that is, for $\Phi(s_1) = \Phi(s_2)$; also there may be elements t_2 in T such that $\Phi(s) = t_2$ never occurs for any s in S. The set S is called the *domain* of the mapping and the set T is called the *codomain*. The set of elements $\Phi(s)$ in T obtained as s runs through S, denoted by $\Phi(S)$ is called the *range* of the mapping. The mapping pictured in figure 4 has range t_1, t_3. If every t in T is the image of at least one element in S, then Φ is said to be a mapping of S *onto* T (see figure 5).

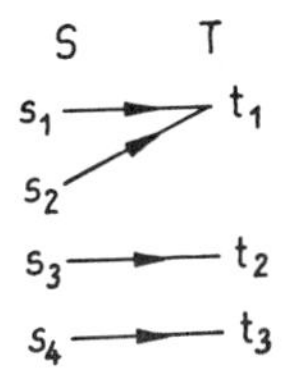

Figure 5

The mapping depicted in figure 4 is not onto. Finally, a mapping Φ is called one–one (1–1) if Φ carries distinct elements into distinct elements, so that $\Phi(s_1) = \Phi(s_2)$ implies $s_1 = s_2$.

The following examples illustrate these concepts.

a) Let S be the integers less than 10 and T be the integers less than 100. The mapping $s \to 10s$ is a 1–1 mapping of S into T whose range is the set $\{\ldots, -30, -20, -10, 0, 10, 20, 30, \ldots, 90\}$.

b) Let S be all positive real numbers and T be all real numbers. The mapping $s \to \log s$ is a 1–1 mapping of S onto T.

c) Let $S = T$ be all complex numbers. The mapping $a + bi \to b$ is a mapping of S into T whose range is the real numbers.

Two mappings $\Phi : S \to T$ and $\Psi : S \to T$ with the same domain S and the same codomain T are called *equal* if they have the same effect upon every point of S, that is, $\Phi = \Psi$ means that $\Phi(s) = \Psi(s)$ for every s in S.

The *product* $\Psi\Phi$ of two transformations is defined as the result of performing them in succession; first Φ, then Ψ, provided that the range of Φ is in the domain of Ψ. In other words, if

$$\Phi : S \to T, \quad \Psi : U \to V,$$

and $\Phi(S)$ is a subset of U, then $\Psi\Phi$ is the transformation of S into V given by $\Psi\Phi(s) = \Psi(\Phi(s))$ which defines the effect of $\Psi\Phi$ upon any point s in S.

In particular, the product of two transformations of S into itself is always defined and is a binary operation of the set of all such transformations.

Multiplication of transformations is associative; i.e. $(\Theta\Psi)\Phi = \Theta(\Psi\Phi)$, whenever the products involved are defined. This follows directly from the definition of the product of two mappings since for every s in S,

$$((\Theta\Psi)\Phi)(s) = (\Theta\Psi)(\Phi(s)) = \Theta(\Psi(\Phi(s)))$$

while

$$(\Theta(\Psi\Phi))(s) = \Theta((\Psi\Phi)(s)) = \Theta(\Psi(\Phi(s))).$$

The definition of equality for mappings proves the associative law.

The permutation π appearing in example 4 can be considered to be a 1–1 transformation Φ of the set $S = \{1, 2, \ldots, n\}$ onto itself by setting $\Phi(i) = i'$. It follows directly from the associative law for mappings that the binary operation of example 4 is associative. The operation defined in example 3 is also associative since each motion can be represented by a permutation of the vertices of the square (see figures 2 and 3).

The importance of the associative law for a set which is closed under an operation is that the *product* $a \circ b \circ c$ of three elements in a specified order is unambiguously defined, since $(a \circ b) \circ c = a \circ (b \circ c)$. The *product* of n elements can be defined as

$$a_1 \circ a_2 \circ \cdots \circ a_n = (a_1 \circ a_2 \circ \cdots \circ a_{n-1}) \circ a_n.$$

This definition is justified by the fact that no brackets are needed in a product, that is

$$a_1 \circ a_2 \circ \cdots \circ a_n = (a_1 \circ a_2 \circ \cdots \circ a_k)$$
$$\circ (a_{k+1} \circ \cdots \circ a_p) \circ \cdots \circ (a_{p+1} \circ a_{p+2} \circ \cdots \circ a_n)$$

no matter how the brackets are arranged*. In particular, for a permutation, we denote $\pi \circ \pi$ by π^2 and define $\pi^n = \pi \circ \pi^{n-1}$.

DEFINITION 4 *An element e is an* identity *if* $e \circ a = a \circ e = a$ *for every a in S.*

In example 1, part 3, the identity for addition is 0 while the identity for multiplication is 1. The identity mapping i of a set S *into* itself is defined by $i(s) = s$ for every s in S. Then for any mapping Φ of S into itself, $i\Phi = \Phi i = \Phi$.

* This result can be proved by induction.

The identity element e (if it exists) for a set S with an operation $\circ$ is unique. Suppose that e' is another identity; then $e \circ e' = e'$ since e is an identity and $e \circ e' = e$ since e' is an identity. Therefore, $e = e'$.

DEFINITION 5 *An element c in S is said to be an* inverse *of an element a in S if $c \circ a = a \circ c = e$, where e is the identity in S.*

In example 1, part 3, the inverse of an element a for addition is $-a$ while its inverse for multiplication is a^{-1} $(a \neq 0)$. The element 0 has no inverse. If Φ is a 1–1 mapping of a set S onto itself, its inverse mapping Ψ is defined by setting $\Psi(t) = s$ where $t = \Phi(s)$ for every s in S. Then $\Psi\Phi(s) = \Psi(\Phi(s)) = s$ and if $s = \Phi(s')$, $\Phi\Psi(s) = \Phi(\Psi(\Phi(s'))) = \Phi(s') = s$. Hence $\Psi\Phi = \Psi\Phi = i$, the identity mapping of S onto itself.

If S is closed under an operation which satisfies the associative law and has an identity, then the inverse c (if it exists) of an element a in S is unique. If c' is another inverse,

$$c = c \circ (a \circ c') = (c \circ a) \circ c' = c'.$$

The symbol $\circ$ for a binary operation is often omitted and the elements are written side by side as in ordinary multiplication. Sometimes the element $+$ is also used to denote the binary operation. The precise notation used is unimportant as long as the symbol is interpreted as an operation (in the proper context) and not thought of as ordinary multiplication or addition.

DEFINITION 6 *A* group *is a set G together with a binary operation obeying the following laws:*

1 *Closure law: If a and b are in G, then ab is in G.*
2 *Associative law: If a, b and c are in G, then $(ab)c = a(bc)$.*
3 *Identity law: There is an element e in G, called the* identity, *such that for every a in G, $ae = ea = a$.*
4 *Inverse law: If a is any element in G, there is an element a^{-1} called the* inverse *of a, such that $aa^{-1} = a^{-1}a = e$.*

If the above notation is used, the binary operation is called (group) *multiplication* and if the symbol $+$ is used, the operation is called (group) *addition*. In additive notation, e is usually denoted by 0 and the inverse of a by $-a$. The associative law becomes

$$(a + b) + c = a + (b + c)$$

If $ab = ba$ in the group, the group is said to be *commutative* or *abelian*. If there are n elements in a group G (n is finite), then the group is said to have *order n*; otherwise the group is of infinite order.

Example 6

1 The numbers $1, -1, \sqrt{-1}, -\sqrt{-1}$ form a group of four elements under multiplication.

2 All rotations about the origin form an infinite group (see example 2).

3 The integers form a group under the operation of addition.

4 All non-zero rational numbers form a group under multiplication.

5 The 8 motions which carry a square into itself form a group (cf. example 3), called the *symmetry group of the square*. Not only the square, but every regular polygon and regular solid has a group of symmetries.

6 The three cube roots of unity form a group under multiplication.

7 The set of all permutations on n symbols form a group when the product of 2 permutations is defined to be the result of their successive performance (see example 4). It is called the *symmetric group* on n symbols. More generally, the set of all 1–1 transformations of any set S onto itself form a group (see example 5).

Many groups are contained in larger groups. The group of even integers under addition is part of the group of all integers under addition. The group of rotations of the square is part of the symmetry group of the square. The group of the eight permutations of the vertices of the square corresponding to the motions of the square (see figures 2 and 3) is part of the symmetric group of all 4! permutations of these vertices. These examples suggest the following concept.

DEFINITION 7 *A subset S of a group G is called a* subgroup *of G if S is itself a group with respect to the binary operation of G.*

The subgroups $\{e\}$ and G are called *trivial* or *improper* subgroups; other subgroups are called *proper* subgroups.

The following theorem, given without proof, characterizes subgroups.

THEOREM 1 *A non-void subset S of a group is a subgroup if and only if*

1 *a and b in S imply that ab is in S.*

2 *If a is in S, then a^{-1} is in S.*

DEFINITION 8 *The* order *of an element g in a group is the least positive integer k such that* $g^k = 1$, *the identity; if no positive power of g equals the identity, g has* order infinity.

DEFINITION 9 *If every element in a group G is equal to some power of a fixed element x in G, the group G is said to be* cyclic. *The element x is said to* generate *the group.*

For example, the cyclic group of all rotations of a square into itself consists of four powers R, R^2, R^3 and $R^4 = I$ of the counterclockwise 90° rotation R. It is also generated by R^3 since $R^2 = (R^3)^2$, $R = (R^3)^3$ and $I = (R^3)^4$. It is not generated by R^2 since R^2 is an element of order 2. Note that $R^3 = R^{-1}$ is a clockwise rotation of 90°.

In general, any negative power of a cyclic group of order n generated by x can be expressed as a positive power—in fact, for $0 < k \leqslant n$

$$x^{-k} = x^{-k}1 = x^{-k}x^n = x^{n-k}.$$

LEMMA 1 *Let G be a cyclic group of order n generated by x. Suppose that the greatest common divisor of the positive integer k* ($k \leqslant n$) *and n is d. Then* x^k *and* x^d *generate the same cyclic subgroup of G; its order is* n/d.

Proof By hypothesis, $k = k_1 d$ and $n = n_1 d$ where k_1 and n_1 are relatively prime. By a well known result* there exist integers a and b such that

$$ak_1 + bn_1 = 1.$$

It follows that

$$x^d = x^{ak_1 d} + x^{bn_1 d} = x^{ak}x^{bn} = x^{ak}$$

since $x^n = 1$. It follows that x^d is an element of the subgroup generated by x^k since it can be expressed as a power of x^k. Therefore, the subgroup generated by x^d is contained in the subgroup generated by x^k. Conversely, the subgroup generated by x^k is contained in the subgroup generated by x^d because $x^k = x^{k_1 d}$. Hence these two subgroups are identical.

To show that this subgroup is of order n/d note that $(x^d)^{n/d} = 1$. Hence the order of x^d is less than or equal to n/d. Suppose $x^{dm} = 1$ where $m < n/d$; then $dm < n$. This implies the order of x cannot be n, which is a contradiction.

---

* G. Birkhoff and S. MacLane, *A Survey of Modern Algebra*, Macmillan Co., New York (1954), p. 18.

COROLLARY *Let G be a cyclic group of order n generated by x. If k and n are relatively prime, then x^k also generates G.*

DEFINITION 10 *A right coset (left coset) of a subgroup S of a group G is any set Sg (gS) of all right-multiples sg (left-multiples gs) of the elements s of S by a fixed element g in G. The number of distinct right cosets* is called the* index *of S in G and will be denoted by* $[G : S]$.

Example 7

1 If G is the group of symmetries of the square, the subgroups $S = \{I, H\}$ has the following four right cosets:

$$SI = \{I, H\} = SH$$

$$SR_1 = \{R_1, HR_1\} = \{R_1, V\} = SV$$

$$SR = \{R, HR\} = \{R, D_2\} = SD_2$$

$$SR_2 = \{R_2, HR_2\} = \{R_2, D_1\} = SD_1$$

Note that each coset has two elements and every element of the group falls into exactly one of the four right cosets. The index of S in G is four.

2 Let G be a cyclic group of order 4 generated by g and let $S = \{e, g^2\}$. Then S decomposes G into two right cosets—namely

$$Se = S \qquad Sg = \{g, g^3\}.$$

These examples illustrate the following general results.

LEMMA 2 *If S is a finite subgroup of a group G, then each right coset Sg of S has exactly as many elements as S.*

Proof Let the element sg in Sg correspond to the element s in S. This mapping is obviously onto. It is also one–one since $s_1g = s_2g$ implies that $s_1 = s_2$. Hence Sg and S have the same number of elements.

LEMMA 3 *Two right cosets Sg and Sh of S are either identical or without common elements.*

Proof Suppose Sg and Sh have an element x in common. Then $x = s_1g = s_2h$ where s_1 and s_2 are in S. Hence Sg contains every element $sh = ss_2^{-1}s_1g$ of Sh. Similarly, Sh contains every element of Sg. Consequently, $Sg = Sh$.

* Left cosets can also be used to define the index (see exercise 8).

LEMMA 4 *A group G is decomposed by means of the distinct cosets of a subgroup S into non-overlapping subsets, each of which has exactly as many elements as S.*

Proof Every element g of G is in some right coset since the coset Sg always contains $g = eg$. the result follows from lemmas 2 and 3.

THEOREM 2 *The order of a finite group G is a multiple of the order of any one of its subgroups S—in fact* $|G| = |S|\,[G:S]$.

Proof From lemma 4 and definition 10, it follows that $|G| = |S|\,[G:S]$.

Exercises

1 Which of the groups in example 6 are abelian?

2 Write out the multiplication tables for the finite groups in example 6.

3 Find the inverse of:

$$\begin{pmatrix}1\,2\,3\,4\,5\,6\\2\,3\,4\,5\,6\,1\end{pmatrix}, \quad \begin{pmatrix}1\,2\,3\,4\,5\,6\,7\\1\,3\,4\,2\,5\,7\,6\end{pmatrix}; \quad \begin{pmatrix}1\,2\,3\,4\,5\,6\,7\,8\\2\,1\,4\,7\,8\,3\,5\,6\end{pmatrix}.$$

4 Write out the multiplication table for the group of order 10 defined by $A^5 = B^2 = 1$, $B^{-1}AB = A^2$.

5 The Klein 4-group is defined by $A^2 = B^2 = 1$, $AB = BA$. Find all subgroups of this group.

6 Prove that if $xx = x$ in a group, then $x = e$.

7 Show that the number of right cosets of any subgroup of a finite group equals the of left cosets.

8 Prove the following:

a) Every group G of prime order p is cyclic.

b) Every element of a finite group has as order a divisor of the order of G.

9 Show that the only groups of order four are the cyclic group of that order and the Klein 4-group.

4 Burnside's lemma

Many people have difficulty doing counting problems such as those in Chapter 1 because distinct objects are often considered equal. If a teacher is interested in knowing the number of families represented by her class then she considers two children to be "equal" if and only if they are siblings. Consider the set of all 3-tuples which can be formed from the first 9 letters of the alphabet (without repetition). If we are interested in the number of 3-combinations, certain 3-tuples are considered to be equal—e.g.,

$$(a, b, c) = (b, c, a) = (a, c, b) = (b, a, c) = (c, a, b) = (c, b, a).$$

As another example, suppose a circular bracelet is formed from one red, one green, one blue and one yellow bead. Two bracelets are considered equal if one can be obtained from the other by a counterclockwise rotation. Distinct bracelets such as those draw below

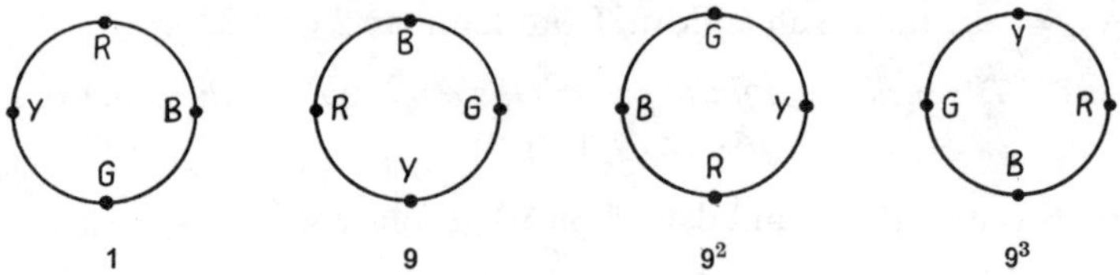

Figure 1

are considered equal.

All objects which are considered equal are placed in a single class called an *equivalence class* and it is the number of such classes which are desired rather than the number of distinct objects. In the first example above, each class represents a surname; in the second, each class represents a 3-combination and in the third, each class represents necklaces which cannot be obtained from each other by a counterclockwise rotation.

The division into classes can be achieved by means of a group of permutations acting on the objects. By definition, *two objects belong to the same class if and only if there is a permutation which carries one into the other*. In the second example of the first paragraph, let G be the group of all permutations acting on 3 objects. Then (x, y, z) is placed in the same class as (x', y', z') if and only if (x', y', z') is a permutation of (x, y, z)—i.e., there exists a g in G such that $g = \begin{pmatrix} x, y, z \\ x', y', z' \end{pmatrix}$. There are 84 classes. In the third example, let each bracelet be represented by 4 equally spaced points on a circle as in the diagram above. Let G be the cyclic group of counterclockwise rotations generated by g, a rotation of 360°/4. If a bracelet such as

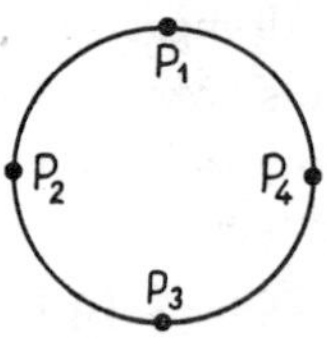

is represented by the symbol $(P_1\ P_2\ P_3\ P_4)$, then $g = \begin{pmatrix} P_1\ P_2\ P_3\ P_4 \\ P_2\ P_3\ P_4\ P_1 \end{pmatrix}$:

Two bracelets $(P_1\ P_2\ P_3\ P_4)$ and $(P_1'\ P_2'\ P_3'\ P_4')$ are considered equivalent and placed in the same class if and only if $g^k(P_i) = P_i'$ ($i = 1, 2, 3, 4$) for some k ($k = 1, 2, 3, 4$) (see figure 1). There are 6 classes,

It is more difficult to construct a group which divides a class of children according to surnames. We illustrate the construction for 3 childıen c_1, c_2, c_3 with the surname Jones and 2 children d_1, d_2 with the surname Smith. Let j be the cycle $(c_1\ c_2\ c_3)$ and s be the cycle $(d_1\ d_2)$ and let G be the group generated by j and s under the assumption that $js = sj$. The reason for this particular choice is that j permutes the c's among themselves and leaves the d's fixed (i.e. $j(d_i) = d_i$ for $i = 1, 2$) while s permutes the d's among themselves and leaves the c's fixed. Since

$$G = \{1, j, j^2, s, js, j^2s\}$$

it is easy to see that G divides the children $\{c_1, c_2, c_3, d_1, d_2\}$ into the two classes $\{c_1, c_2, c_3\}$ and $\{d_1, d_2\}$. In more detail, $j(c_1) = c_2$, $j(c_2) = c_3$, $j(c_3) = c_1$ and so the c's are all in the same class, while $s(d_1) = d_2$ and $s(d_2) = d_1$ and so the d's are all in the same class. There is no element of G which carries a d into a c or vice versa.

In the above examples, it was easy to find the number of equivalence classes by merely counting. However for more complex problems, this is impractical. Fortunately, Burnside's lemma gives an analytic formula for the number of equivalence classes. Before proving this lemma, we will study the way to divide an arbitrary set into subsets such that every element of the original set belongs to exactly one of the subsets.

DEFINITION 1 *Let G be a group of permutations of a set S. The element s_2 in S is said to be* equivalent *(with respect to G) to s_1 in S if there exists an element g in G such that $gs_1 = s_2$. The* equivalence class *of s_1 (with respect to G), denoted by $C(s_1)$ consists of all s in S which are equivalent to s_1.*

Note that if s is in $C(s_1)$, then $gs_1 = s$ for some g in G. In this situation, $s_1 = g^{-1}s$ and so s_1 is equivalent to s. Hence s is equivalent to s_1 if and only if s_1 is equivalent to s. Therefore we will simply say that s_1 and s are equivalent.

LEMMA 1 *Every element in S belongs to some equivalence class.*

In fact, s is in $C(s)$ since $s = es$ where e is the identity in G.

LEMMA 2 *The elements s_1 and s_2 are equivalent if and only if $C(s_1) = C(s_2)$.*

Proof If the equivalence class $C(s_2)$ is identical to $C(s_1)$ then s_1 is in $C(s_2)$ and so, by definition, s_1 is equivalent to s_2.

Conversely, suppose that s_1 and s_2 are equivalent. Then there exists a g in G such that $s_1 = gs_2$. Let s be an element of $C(s_1)$; then $s = hs_1$. Hence $s = hs_1 = hgs_2$ which implies that s is in $C(s_2)$. Therefore every element of $C(s_1)$ is also an element of $C(s_2)$. An analogous computation shows that every element of $C(s_2)$ is also in $C(s_1)$ and so $C(s_1) = C(s_2)$.

COROLLARY *Two equivalence classes are either identical or without common elements.*

Proof Suppose that s is in $C(s_1)$ and $C(s_2)$. Then s and s_1 are equivalent and so $C(s) = C(s_1)$ by the lemma. Similarly, $C(s) = C(s_2)$ which proves the result.

Lemma 1 and the corollary of lemma 2 imply the following result.

THEOREM 1 *Let G be a group of permutations of a set S. Then S can be decomposed into non-overlapping subsets which are the equivalence classes with respect to G.**

The converse result also holds. If S is decomposed into non-overlapping subsets, then there is a group G for which these subsets are equivalence classes. We omit the proof of this result. The proof is a generalization of the one given in the example on the decomposition according to surnames appearing on page 183.

The following results will be required for the proof of Burnside's lemma.

LEMMA 3 *The set of all elements g in G which leave s fixed (i.e. $g(s) = s$) form a subgroup $H(s)$.*

Proof If g and h are in $H(s)$, then gh is in $H(s)$ since $gh(s) = g(s) = s$. If g is in $H(s)$, then g^{-1} is in $H(s)$ since $g(s) = s$ implies that $s = g^{-1}(s)$. Therefore, $H(s)$ is a subgroup by theorem 3.1.

LEMMA 4 *The number of elements in $C(s)$ is $[G : H(s)]$.*

Proof The set $C(s)$ is comprised of all the different elements $g(s)$ obtained by letting g run through G. $g_1(s) = g_2(s)$ if and only if $g_2^{-1}g_1(s) = s$ or, equivalently, $g_2^{-1}g_1$ is in $H(s)$; this is equivalent to g_1 being in the left coset

* The advanced reader will note that G induces an equivalence relation on S and theorem 1 follows from a more general result. Consult p. 155 of the reference cited in footnote 3 of § 3.

g_2H. Hence $g_1(s) = g_2(s)$ if and only if g_1 and g_2 are in the same coset. It follows that the number of elements in $C(s)$ is the number of different cosets which is just $[G : H(s)]$.

LEMMA 5 *If* $s_1 = g(s)$, *then*

$$H(s_1) = gH(s)\,g^{-1}. \tag{1}$$

Proof Let k be an element of $H(s_1)$—that is, $k(s_1) = s_1$. Substituting $g(s)$ for s_1 in the last equation yields $kg(s) = g(s)$ or $g^{-1}kg(s) = s$. Therefore, $g^{-1}kg$ is in $H(s)$ or k is in $gH(s)\,g^{-1}$. Thus every element of $H(s_1)$ is an element of $gH(s)\,g^{-1}$.

Conversely, let k be in $gH(s)\,g^{-1}$; then $k = ghg^{-1}$, where h is an element of $H(s)$. This implies that

$$k(s_1) = ghg^{-1}(s_1) = gh(s) = g(s) = s_1$$

and so k is in $H(s_1)$. Since every element of $gH(s)\,g^{-1}$ is an element of $H(s_1)$, the result follows.

COROLLARY *If* s_1 *is in* $C(s)$, *then* $H(s)$ *and* $H(s_1)$ *have the same number of elements.*

The corollary follows from the fact that the mapping of $H(s)$ into $H(s_1)$ defined by $h \to ghg^{-1}$ is obviously 1–1 and onto by (1).

We are now prepared to prove Burnside's lemma.

THEOREM 2 *Let* G *be a finite group of permutations of a finite set* S. *The number of equivalence classes established in* S *by* K *is*

$$|G|^{-1} \sum_{g \text{ in } G} I(g)$$

where $|G|$ *denotes the number of elements in* G *and* $I(g)$ *is the number of elements* s *in* S *left fixed by* g (*that is, for which* $g(s) = s$).

Proof Let $\delta(g(s), s) = 1$ if $g(s) = s$ and 0 if $g(s) \neq s$; and let C denote an equivalence class in S. Then

$$\begin{aligned} |G|^{-1} \sum_{g \text{ in } G} I(g) &= |G|^{-1} \sum_{g \text{ in } G} \sum_{s \text{ in } S} \delta(g(s), s) \\ &= |G|^{-1} \sum_{s \text{ in } S} \sum_{g \text{ in } G} \delta(g(s), s) \\ &= |G|^{-1} \sum_{s \text{ in } S} (\text{no. of permutations which leave } s \text{ fixed}) \\ &= |G|^{-1} \sum_{C \text{ in } S} \sum_{s \text{ in } C} (\text{no. of permutations which leave } s \text{ fixed}) \quad (2) \end{aligned}$$

where the last equality is a consequence of theorem 1.

Let $H(s)$ denote the group of permutations which leave s fixed. By the corollary of lemma 5, $H(s) = H(s_1)$ for any s_1 in $C(s)$. Hence

$$\sum_{s \text{ in } C} (\text{no. of permutations which leave } s \text{ fixed})$$
$$= (\text{no. of } s \text{ in } C)\,|H(s)| = [G : H(s)]\,|H(s)| = |G| \tag{3}$$

by lemma 4 and theorem 3.2. Substituting from (3) into (2) yields

$$|G|^{-1} \sum_{g \text{ in } G} I(g) = |G|^{-1} \sum_{C \text{ in } S} |G| = \text{no. of equivalence classes.}$$

The following examples illustrate how to use Burnside's lemma.

Example 1

1 Let S be the set of all 3-tuples which can be formed from the first nine letters of the alphabet and let G be the group consisting of the identity permutation $e = \begin{pmatrix} x, y, z \\ x, y, z \end{pmatrix}$. Then there is 1 element in each class and the number of equivalence classes equals the number of 3-permutations. By Burnside's lemma, this number is $I(e) = 9 \cdot 8 \cdot 7$ since every element is left fixed by e.

2 Let S be as in 1. Then, as discussed previously, to find the number of 3-combinations we take G to be the symmetric group of order 3 and find the number of equivalence classes. By Burnside's lemma, this number is $\dfrac{9 \cdot 8 \cdot 7}{1 \cdot 2 \cdot 3}$ since $I(e) = 9 \cdot 8 \cdot 7$, $|G| = 6$ while $I(g) = 6$ for $g \neq e$ and g in G.

3 Now let us apply Burnside's lemma to calculate the number of surnames in the example appearing on page 183. Since $|G| = 6$ and $I(1) = 5$, $I(j) = I(j^2) = 2$, $I(s) = 3$ and $I(js) = I(j^2s) = 0$, Burnside's lemma yields 2 equivalence classes.

4 The six faces of a cube are to be painted with six different colors, each face a different color. In how many distinguishable ways can this be done? Two cubes are regarded as similar if one of them can be rotated in such a fashion that we can no longer see their differences.

In this problem, G is the group of 24 rotations acting on the set S of all cubes painted 6 different colors (see example 5.2). If a cube having each face a different color is operated on by any rotation except the identity, it cannot be left fixed. Therefore, $I(g) = 0$ for $g \neq e$. The number of ways of painting a cube without regard to symmetric coloring is $6! = 720$. By Burnside's lemma, the number of distinguishable colorings is $\frac{1}{24}(720) = 30$.

When G is a cyclic group, Burnside's lemma can be simplified by means of the following result.

LEMMA 6 *If g and h are elements of the group of permutations G of a finite set S which generate the same cyclic subgroup, then $I(g) = I(h)$.*

Proof Since g and h generate the same cyclic subgroup, there exist positive integers a and b such that $g^a = h$ and $h^b = g$. Let s be any element left fixed by g. Then

$$h(s) = g^a(s) = g^{a-1}\, g(s) = g^{a-1}(s) = \cdots = s$$

and so s is also left fixed by h. In other words, every element of $I(g)$ is also an element of $I(h)$. Reversing the roles of g and h, we can deduce that $I(g) = I(h)$.

THEOREM 3 *If G is a cyclic group of order n generated by the element g, then the number of equivalence classes in a set S under the permutations of G is*

$$n^{-1} \sum_{d|n} I(g^d)\phi\,(n/d)$$

where ϕ is Euler's function and summation is over the divisors d of n.

Proof By lemma 3.1, g^k and g^d generate the same cyclic subgroup, where $d = (k, n)$, the greatest common divisor of k and n. The number of solutions of $(k, n) = d$ is the same as the number of solutions of $(k/d, n/d) = 1$. By definition of Euler's function (see exercise 4.1.3) this number is $\phi\,(n/d)$. Since $I(g^d) = I(g^k)$, $I(g^d)$ is counted with a multiplicity of $\phi\,(n/d)$ and so Burnside's lemma yields

$$|G|^{-1} \sum_{g \text{ in } G} I(g) = n^{-1} \sum_{d|n} I(g^d)\phi\,(n/d).$$

Example 2 Find the number of distinct circular necklaces with n beads on a string using c colors of beads if two necklaces which can be rotated into each other are regarded as equivalent.

In this case, S is the set of n equally spaced points on a circle and G the cyclic group whose elements g_k rotate the elements of S $2\pi k/n$ degrees. Since g_d rotates d consecutive beads to the next block of d consecutive beads, a necklace can be invariant under g_d if and only if each block of d beads has the same color arrangement. In other words, the color of the necklace is completely determined once the color of d consecutive beads is specified. This can be done in c^d ways and so $I(g^d) = c^d$. Hence the number of distinct necklaces is

$$\frac{1}{n} \sum_{d|n} \phi\left(\frac{n}{d}\right) c^d$$

by theorem 3. If n is a prime number p, this number reduces to p^{-1} $[c^p + (p-1)c]$.

Finally, we give an alternate proof of Burnside's lemma based on Möbius inversion. This further illustrates the fact that many seemingly unrelated combinatorial problems are in reality interconnected via the Möbius function. This proof requires the following lemmas.

LEMMA 7 *Let G be a group of permutations acting on a finite set S and K, L be subgroups of G. Let $I(K)$ denote the number of elements s in S left fixed by every k in K and $I^*(L)$ denote the number of elements s such that $H(s) = L$. Then*

$$I(K) = \sum_{L \geqslant K} I^*(L) \tag{4}$$

where the summation is over all subgroups L containing K according to the notation introduced in example 1.2.

Proof Suppose an element s is counted in the left side of (4). There exists a unique subgroup $L = H(s)$ consisting of all elements which leave s fixed. Then $L \geqslant K$ and so s is counted once in the right side of (4). Conversely, if s is counted in the right side of (4), it is left fixed by every element of some subgroup L and so is left fixed by K; accordingly, s must be counted in $I(K)$.

COROLLARY

$$I^*(K) = \sum_{L \geqslant K} I(L)\,\mu\,(K, L) \tag{5}$$

The corollary follows immediately upon applying Möbius inversion to (4) (see exercises 2.2 part (b)).

LEMMA 8 *Let K and L be subgroups of a finite group G. Define a function n on the subgroups by $n(L)$ = number of generators of L if L is cyclic and $n(L) = 0$ otherwise. Then*

$$|K| = \sum_{L \leqslant K} n(L). \tag{6}$$

Proof Divide K into nonoverlapping classes by placing two elements of K in the same class provided they generate the same cyclic subgroup. Since every k in K generates a cyclic subgroup, every k in K lies in one of these classes. A unique cyclic subgroup of K corresponds to each class and conversely, every cyclic subgroup L of K gives rise to a class—namely the class of all elements which generate L. Since the number of elements in a class corresponding to a cyclic subgroup L is $n(L)$, (6) follows by noting that $n(L) = 0$ if L is not cyclic.

COROLLARY

$$n(K) = \sum_{L \leqslant K} |L| \, \mu(L, K) \tag{7}$$

The corollary follows immediately from (6) by applying Möbius inversion (theorem 2.1).

Alternate proof of Burnside's lemma If s is in S and $m = |S|$, then $|C(s)| = j$ for some $j \, (1 \leqslant j \leqslant m)$. Therefore, the total number of equivalence classes is

$$\sum_{j=1}^{m} \sum_{\{C(s):|C(s)=j\}} 1 \tag{8}$$

where $\{C(s): |C(s)| = j\}$ denotes that the summation is over all $C(s)$ which contain j elements. Using the identity

$$j^{-1} \sum_{\{s:|C(s)|=j\}} 1 = \sum_{\{C(s):|C(s)|=j\}} 1,$$

where $\{s: |C(s)| = j\}$ denotes that the summation is over all s for which $|C(s)| = j$, (8) reduces to

$$\sum_{j=1}^{m} j^{-1} \sum_{\{s:|C(s)=j\}} 1. \tag{9}$$

(9) can then be rewritten as

$$\sum_{j=1}^{m} j^{-1} \sum_{K \leqslant G} \sum_{\{s:|C(s)|=j,\, H(s)=K\}} 1 \tag{10}$$

where $\{s: |C(s)| = j, H(s) = K\}$ denotes that the summation is over all s for which $|C(s)| = j$ and $H(s) = K$. By lemma 4, $j = [G:K] = |G|/|K|$ (by theorem 3.2) which when substituted into (10) yields

$$\sum_{j=1}^{m} \frac{|K|}{|G|} \sum_{K \leqslant G} \sum_{\{s:|C(s)|=j,\, H(s)=K\}} 1 = \sum_{K \leqslant G} \frac{|K|}{|G|} \left(\sum_{j=1}^{m} \sum_{\{s:|C(s)|=j,\, H(s)=K\}} 1 \right) \tag{11}$$

The expression in brackets in the right side of (11) is just $I^*(K)$, defined in lemma 7, and so (11) reduces to

$$|G|^{-1} \sum_{K \leqslant G} |K|\, I^*(K) = |G|^{-1} \sum_{K \leqslant G} |K| \sum_{L \geqslant K} I(L)\, \mu\,(K, L), \tag{12}$$

the last equality being a consequence of the corollary of lemma 7. Since $\mu\,(K, L) = 0$ for $L \not\geqslant K$,

$$\begin{aligned} |G|^{-1} \sum_{K \leqslant G} |K| \sum_{L \geqslant K} I(L)\, \mu\,(K, L) &= |G|^{-1} \sum_{K \leqslant G} |K| \sum_{L \leqslant G} I(L)\, \mu\,(K, L) \\ &= |G|^{-1} \sum_{L \leqslant G} I(L) \sum_{K \leqslant G} |K|\, \mu\,(K, L) \\ &= |G|^{-1} \sum_{L \leqslant G} I(L) \sum_{K \leqslant L} |K|\, \mu\,(K, L) \\ &= |G|^{-1} \sum_{L \leqslant G} I(L)\, n(L) \quad \text{(by the corollary of lemma 8)} \\ &= |G|^{-1} \sum_{g \text{ in } G} I(\langle g \rangle), \end{aligned}$$

where $\langle g \rangle$ denotes the cyclic subgroup generated by g. Since $I(\langle g \rangle) = I(g)$, the proof of Burnside's lemma is complete.

Exercises

1 In how many different ways (cf. example 1, part 4) can a cube be colored with 3 black faces and 3 red faces?

2 Show that the number of distinct ways that the faces of a cube can be painted with c colors, where not all colors need be used, is

$$\frac{1}{24}\,(c^6 + 3c^4 + 12c^3 + 8c^2).$$

3 Suppose that each vertex of a cube is painted black or white. Show that there are 23 differently painted cubes.

4 Let G be a permutation-group acting on a finite set S. Let g and h be elements of G with $g = k^{-1}hk$ for some fixed k in G. Prove that $I(g) = I(h)$.

5 Work example 1.4.6 by using Burnside's lemma.

6 Let W be a function (weight) defined on the equivalence classes. Prove the following weighted Burnside's lemma:

$$\sum_{C(s)} W(C(s)) = |G|^{-1} \sum_{g \text{ in } G} I(\langle g \rangle).$$

7 All four digit numbers are to be written on cards, one per card with leading zeros filled in. If a card is read either right side up or upside down show that $\frac{1}{2}(10^4 + 10^4 - 5^4 + 3\cdot 5)$ cards are required. [Hint: The same card can be used for 0001 and 1000; 8016 and 9108 etc. Let $G = \{g_1, g_2\}$ where g_1 is the identity mapping and g_2 is defined as follows: if a number n is m when read upside down, let $g_2(n) = m$; if n cannot be read upside down, $g_2(n) = n$— e.g., $g_2(0001) = 1000$. $I(g_2) = 10^4 - 5^4 + 3 \cdot 5$].

5 The cycle index

The next section deals with Polya's theorem, often called the fundamental theorem in enumerative combinatorial analysis. The understanding of this result requires the concept of the cyclic index of a permutation group. We begin by defining a cycle of a permutation.

DEFINITION 1 *A permutation*

$$\pi = \begin{pmatrix} s_1 & s_2 & s_3 & \cdots & s_n \\ s_1' & s_2' & s_3' & \cdots & s_n' \end{pmatrix}$$

on n distinct elements $s_1, s_2, \ldots, s_n$ is called a cycle (*or* cyclic permutation) *if $s_1' = s_2, s_2' = s_3, \ldots, s_{n-1}' = s_n, s_n' = s_1$.*

Thus, for example, $\begin{pmatrix} 1\,2\,3\,4 \\ 2\,3\,4\,1 \end{pmatrix}$ and $\begin{pmatrix} a\,b\,c\,d\,e \\ b\,c\,d\,e\,a \end{pmatrix}$ are cycles. They are usually written more simply as $(1\,2\,3\,4)$ and $(a\,b\,c\,d\,e)$, respectively, where each symbol is followed by its image. If we are using a permutation on 8 elements, $(s_1\,s_5\,s_8)$ is a cycle representing the permutation

$$\begin{pmatrix} s_1 & s_2 & s_3 & s_4 & s_5 & s_6 & s_7 & s_8 \\ s_5 & s_2 & s_3 & s_4 & s_8 & s_6 & s_7 & s_1 \end{pmatrix} \quad \text{or} \quad \begin{pmatrix} s_1 & s_5 & s_8 \\ s_5 & s_8 & s_1 \end{pmatrix}$$

it being conventional that any symbols not appearing in a permutation remain unaltered by it.

THEOREM 1 *Any permutation π can be written as a product of cycles, no two of which alter the same symbol.*

Proof Choose any letter s_i and find $s_i' = \pi(s_i)$; then let $(s_i')' = s_i'' = \pi^2(s_i)$ and so on. Since there are at most n symbols, $s_i \doteq \pi^r(s_i) \equiv s_i^{(r)}$ is ultimately reached and we have a cycle

$$(s_i s_i' s_i'' \cdots s_i^{(r-1)}).$$

If $r = n$, the permutation is composed of one cycle of length n; if $r < n$ there is an s_j not in the cycle already obtained. Starting from it, we build up a cycle

$$(s_j s_j' s_j'' \cdots s_j^{(k-1)}).$$

Continuing if necessary, we finally end up with the given permutation written as a product of cycles.

The following example illustrates the procedure used in the above theorem. Suppose that

$$\pi = \begin{pmatrix} 1\,2\,3\,4\,5\,6\,7\,8\,9 \\ 3\,4\,9\,7\,6\,8\,1\,5\,2 \end{pmatrix}$$

and the element 4 is chosen. Since $4 \to \pi(4) = 7 \to \pi^2(4) = 1 \to 3 \to 9 \to 2 \to 4$ we have the cycle (4 7 1 3 9 2). Next, selecting an element not appearing in the cycle, 5, for example, and noting that $5 \to 6 \to 8 \to 5$ yields the cycle (5 6 8). All the numbers from 1 to 9 now appear in one of the two cycles. Hence

$$\pi = (4\,7\,1\,3\,9\,2)\,(5\,6\,8) = (5\,6\,8)\,(4\,7\,1\,3\,9\,2). \tag{1}$$

The identity permutation e has cyclic decomposition (1) (2) (3) ··· (9). As further illustrations,

$$\pi_2 = \begin{pmatrix} 1\,2\,3\,4\,5\,6 \\ 2\,1\,4\,3\,6\,5 \end{pmatrix} = (1\,2)\,(3\,4)\,(5\,6); \tag{2}$$

$$\pi_3 = \begin{pmatrix} 1\,2\,3\,4\,5 \\ 2\,5\,4\,3\,1 \end{pmatrix} = (1\,2\,5)\,(3\,4). \tag{3}$$

Note that the representation of a cycle is not unique; for instance

$$(1\,2\,3) = (2\,3\,1) = (3\,1\,2).$$

Moreover, if cycles are disjoint (there is no element common to any two of the cycles), their product is independent of the order in which they are written; for example,

$$\pi_2 = (1\,2)\,(3\,4)\,(5\,6) = (3\,4)\,(1\,2)\,(5\,6) = (5\,6)\,(1\,2)\,(3\,4) = \cdots$$

Hence there are several ways of representing a permutation as a product of disjoint cycles. What is unique is the number of cycles of length r ($r = 1, 2, \ldots, n$) where n is the total number of elements in the set. In order to prove this, assume that the assertion is false. Suppose that the permutation has two different cyclic decompositions as described in theorem 1. Then there exists some element s which is part of a cycle of length j and also part of a cycle of length k, where $k \neq j$. This is impossible and so the uniqueness of the number of cycles of length r is proved. Hence the following definition is sensible.

DEFINITION 2 *Let π be a permutation acting on a set S of n elements. If π can be written as a product of k_1 disjoint cycles of length* **1**, *k_2 disjoint cycles of length* **2**, *…, k_n disjoint cycles of length n, then π is said to be of* type $\{k_1, k_2, \ldots, k_n\}$.

Note that $k_1 + 2k_2 + \cdots + nk_n = n$, since the sum of the lengths of the cycles is the total number of elements in S.

The types of the permutations π, e, π_1, π_2 (see (1), (2) and (3)) are

$$(0, 0, 1, 0, 0, 1, 0, 0, 0), \quad (9, 0, 0, 0, 0, 0, 0, 0, 0), \quad (0, 3, 0, 0, 0, 0),$$
$$(0, 1, 1, 0, 0),$$

respectively.

We are now prepared to define the cycle index of a permutation group.

DEFINITION 3 *Let G be a group whose elements are permutations of a set S of n elements. If g in G is of type $(k_1(g), k_2(g), \ldots, k_n(g))$ form the product $x_1^{k_1(g)}x_2^{k_2(g)} \cdots x_n^{k_n(g)}$ where $x_1, \ldots, x_n$ are n variables. Taking the sum of these terms and dividing by the number of elements $|G|$ in G the polynomial so obtained,*

$$P_G(x_1, x_2, \ldots, x_n) = |G|^{-1} \sum_{g \text{ in } G} x_1^{k_1(g)}x_2^{k_2(g)} \cdots x_n^{k_n(g)},$$

is called the cycle index *of G.*

The following examples illustrate the calculations involved in finding the cycle index of a group G.

Example 1 Let G consist of the identity permutation only. Since the identity permutation is of type $(n, 0, \ldots, 0)$, $P_G = x_1^n$.

Note that P_G depends on n. Hence the cycle index of G depends not only on the structure of G as an abstract group but also on the interpretation of its elements as permutations of a set S. Example 2 and exercises 1 and 2 further illustrate this point.

Example 2 Let S be the set of 8 vertices of a cube and let G be the group of permutations of S that can be produced by rotations of the cube onto itself. The $8 \times 3 = 24$ such rotations can be grouped into the following five categories:

a) The identity.

b) Three 180° rotations around lines connecting the centers of opposite faces.

c) Six 90° rotations around lines connecting the centers of opposite faces.

d) Six 180° rotations around lines connecting the midpoint of opposite edges.

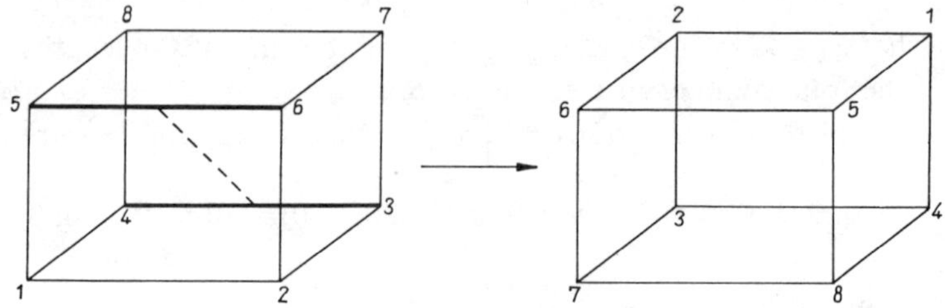

e) Eight 120° rotations around lines connecting opposite vertices. (In the diagram below, the rotation is counterclockwise about the axis through 6 and 4. This axis passes through the centroid of the triangle whose vertices are 5, 2, and 7.)

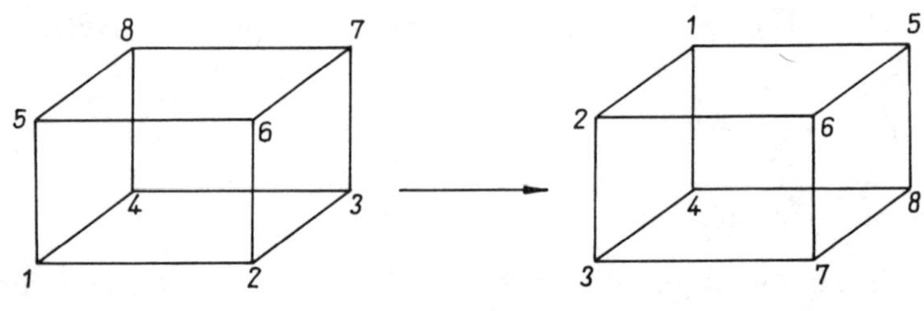

There are eight cycles of length 1 in (a). The permutations in the remaining categories can be decomposed as follows: in (b)—into four cycles of length 2; in (c)—into two cycles of length 4; in (d)—into four cycles of length 2; in (e)—into two cycles of length 1 and two of length 3. Hence, the cycle index is $P_G = \frac{1}{24}(x_1^8 + 3x_2^4 + 6x_4^2 + 6x_2^4 + 8x_1^2x_3^2) = \frac{1}{24}(x_1^8 + 9x_2^4 + 6x_4^2 + 8x_1^2x_3^2)$.

Example 3 Let S be a set of n elements and let G be the symmetric group of degree n. A typical element of G is of type $\{k_1, k_2, \ldots, k_n\}$ where $k_1 + 2k_2 + 3k_3 + \cdots + nk_n = n$. To find the cycle index of G, we must first find x, the number of elements of G of the above type (the k_i are fixed). A cycle of length r can be represented in r different ways by cyclically permuting the elements composing the cycle. Hence the elements in k_r cycles of length r can be rearranged in r^{kr} different ways without the product of these cycles being altered. The product still remains unaltered if the k_r factors are written in a different order; there are $k_r!$ different ways that this can be done. Hence an element of type $\{k_1, k_2, \ldots, k_n\}$ can be represented in $k_1!\,1^{k_1}k_2!\,2^{k_2}\cdots k_n!\,n^{k_n}$ different ways as the product of cycles, and so x elements can be represented in

$$(k_1!\,1^{k_1}k_2!\,2^{k_2}\cdots k_n!\,n^{k_n})\,x \tag{4}$$

ways. We will find the value of x by regarding each element as an n-permutation (in the sense of chapter 1) by removing the brackets which decompose the elements of S into cycles. Under this interpretation, (4) is just the number of n permutations, $n!$. Thus

$$x = n!\,(k_1!\,k_2!\,2^{k_2}\cdots k_n!\,n^{k_n})^{-1}.$$

Accordingly, the cycle index of G is

$$P_G = \sum (k_1!\,k_2!\,2^{k_2}\cdots k_n!\,n^{k_n})\,x_1^{k_1}\,x_2^{k_2}\cdots x_n^{k_n}$$

where the summation is over all n-tuples $(k_1, k_2, \ldots, k_n)$ of nonnegative integers k_i which satisfy

$$k_1 + 2k_2 + 3k_3 + \cdots + nk_n = n.$$

Example 4 Let G be a cyclic group of permutations of order n generated by π acting on a set S of n elements. Let $g_k = \pi^k$ $(k = 0, 1, 2, \ldots, n-1)$ and let (s, t) denote the greatest divisor of s and t. By lemma 3.1, the same cyclic subgroup is generated by g_k and $g_{(k,n)}$. In other words, if $(k, n) = (j, n) = d$, then g_j, g_k and g_d all generate the same cyclic subgroup; moreover, they are all of the same type. For if s is any element in S then it belongs to the cycle obtained from

$$s \to g_d s \to g_d^2 s \to \cdots \to g_d^{n/d} s = s.$$

It follows that g_d is the product of d cycles of length n/d. To calculate the cycle index we must find the number of solutions of $(k, n) = d$ or $(k/d, n/d) = 1$, since for all such k the corresponding g_k are of the same type. The number of solutions of $(k/d, n/d) = 1$ is, by the definition of

Euler's totient function (see exercise 4.1.3) $\phi(n/d)$. Hence the cycle index of G, $C_n(x_1, x_2, \ldots, x_n)$ is

$$C_n(x_1, x_2, \ldots, x_n) = \frac{1}{n} \sum_{d|n} \phi(n/d)\, x_{n/d}^{d} = \frac{1}{n} \sum_{t|n} \phi(t)\, x_t^{n/t}. \tag{5}$$

Example 5 The dihedral group of degree n is the group of symmetries of a regular n-gon (n sided polygon) and is generated by a rotation and a reflection; it is naturally associated with circular ordering when the two directions for traversing the circle in reading an arrangement of its points are regarded as equivalent. We will show that the cycle index, D_n, of the dihedral group is given by

$$D_n(x_1, \ldots, x_n) = \tfrac{1}{2}\left(C_n(x_1, \ldots, x_n) + x_1 x_2^{m}\right) \quad (n = 2m + 1)$$

$$= \tfrac{1}{2}\left(C_n(x_1, \ldots, x_n) + \tfrac{1}{2}\left(x_1^2 x_2^{m-1} + x_2^{m}\right)\right) \quad (n = 2m), \tag{6}$$

where C_n is defined in (5).

We will only prove (6) for $n = 2m$; the case $n = 2m + 1$ is left as an exercise for the reader. We begin by describing the structure of the dihedral group G. A regular n-gon is depicted in the diagram below.

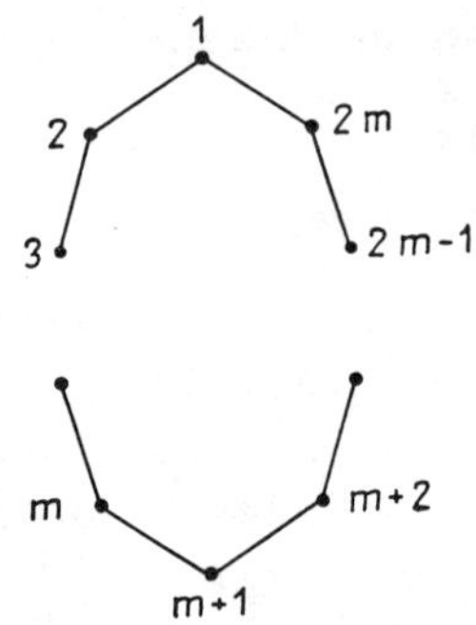

Let c be a counterclockwise rotation of the n-gon about its center through $\left(\frac{360}{n}\right)^{\circ}$; then $c(i) = i + 1$ $(i = 1, 2, \ldots, n - 1)$ and $c(n) = 1$. Let r be a reflection in the diagonal joining the points 1 and $m + 1$; then $r(1) = 1$ and $r(k) = 2m + 2 - k$ for $k = 2, 3, \ldots, 2m$. It is not difficult to verify that

$$c^n = r^2 = 1; \quad rc = c^{n-1} r. \tag{7}$$

The last relations imply that

$$rc^2 = (rc)\,c = c^{n-1}\,(rc) = c^{n-1}\,(c^{n-1}r) = c^{2n-2}r = c^{n-2}r,$$
$$rc^3 = (rc^2)\,c = c^{n-2}\,(rc) = c^{n-2}\,(c^{n-1}r) = c^{2n-3}r = c^{n-3}r$$

and more generally that

$$rc^k = c^{n-k}r \quad (k = 1, 2, \ldots, n). \tag{8}$$

Relations (7) and (8) imply that the dihedral group has $2n$ elements—more specifically,

$$G = \{1, c, c^2, \ldots, c^{n-1}, r, rc, \cdots, rc^{n-1}\}.$$

An element rc^k ($k = 0, 1, \ldots, n-1$) is of order 2 since it follows from (8) and (7) that

$$(rc^k)\,(rc^k) = rc^k c^{n-k} r = r^2 = 1.$$

To find the cycle index of G we need only find the type of elements of the form rc^i ($i = 0, \ldots, n-1$) since the type of elements of the form c^i is known (see example 4). The cyclic decomposition of rc^i consists only of cycles of length 1 or cycles of length 2 since rc^i is of order 2. Since

$$\begin{aligned} c^i(j) &= j + i & (j + i \leqslant 2m)\\ &= j + i - 2m & (j + i > 2m)\end{aligned}$$

it follows that

$$\begin{aligned} rc^i\,(j) &= 2m + 2 - j - i & (1 < j + i \leqslant 2m)\\ &= 4m + 2 - j - i & (j + i > 2m)\end{aligned}$$

Hence $rc^i(j) = j$ if and only if

$$2m + 2 - j - i = j \quad \text{or} \quad 4m + 2 - j - i = j.$$

These last two equations have a solution only if i is even, say $i = 2k$. The solutions are $j = m + 1 - k$ and $j = 2m + 1 - k$ ($k = 1, 2 \ldots, m-1$); $j = m + 1$ and $j = 1$ ($k = 0$). Hence rc^{2k} ($k = 0, 1, \ldots, m-1$) contains two cycles of length 1 and $m-2$ cycles of length 2 while rc^{2k+1} produces m cycles of length 2. Using (5) it follows that the cycle index of G is

$$D_n = \frac{1}{2}\,(C_n\,(x_1, \ldots, x_n) + \frac{1}{2n}\,(mx_1^2x_2^{m-1} + mx_2^m)$$

which reduces to (6) for $n = 2m$.

Example 6 Let G, H be groups of permutations of sets S, T respectively. Assume that S and T have no elements in common. Form a set U as follows: u is in U if and only if u is in S or u is in T. Each g in G and h in H can be used to define a permutation of U by setting

$$u \to g(u) \quad \text{if } u \text{ is in } S; \qquad u \to h(u) \quad \text{if } u \text{ is in } T.$$

This permutation of U will be denoted by gxh. Defining the product of two permutations g_1xh_1 and g_2xh_2 by

$$(g_1xh_1)(g_2xh_2) = g_1g_2xh_1h_2$$

it is not difficult to verify that these permutations form a group. This group is called the *direct product* of G and H and is denoted by GxH. Its order is the product of the orders of G and H.

If g in G is of type $\{k_1, k_2, \ldots\}$ and h in H is of type $\{j_1, j_2, \ldots\}$, then gxh is of type $\{k_1 + j_1, k_2 + j_2, \ldots\}$, since each cycle in U lies entirely in S or entirely in T. Therefore, the term of the cycle index of GxH corresponding to the element gxh is equal to the product of the term in P_G corresponding to g and the term in P_H corresponding to h. Summing all such terms as g varies in G and h varies in H and dividing by the order of GxH yields, by definition, P_{GxH}. On the other hand, it is also equal to

$$|GxH|^{-1}\left(\sum_{g \text{ in } G} x_1^{k_1(g)} x_2^{k_2(g)} \cdots\right)\left(\sum_{h \text{ in } H} x_1^{j_1(h)} x_2^{j_2(h)} \cdots\right) = P_G \cdot P_H$$

since $|GxH| = |G| \cdot |H|$. Hence

$$P_{GxH} = P_G \cdot P_H.$$

Exercises

1 Write the following as a product of disjoint cycles:

$$\begin{pmatrix}1\,2\,3\,4\,5\,6\\2\,1\,5\,6\,3\,4\end{pmatrix}, \quad \begin{pmatrix}1\,2\,3\,4\,5\,6\,7\,8\\2\,7\,6\,4\,8\,1\,3\,5\end{pmatrix}, \quad \begin{pmatrix}1\,2\,3\,4\,5\\5\,4\,3\,2\,1\end{pmatrix}, \quad \begin{pmatrix}1\,2\,3\,4\,5\,6\,7\,8\,9\\2\,7\,1\,6\,4\,9\,8\,5\,3\end{pmatrix}$$

2 Let S be the edges of a cube and G be the group of rotations described in example 2. Show that

$$P_G = \tfrac{1}{24}(x_1^{12} + 3x_2^6 + 6x_4^3 + 6x_1^2x_2^5 + 8x_3^4.$$

3 Let S be the faces of a cube and G be the group of rotations described in example 2˙ Show that

$$P_G = \tfrac{1}{24}(x_1^6 + 3x_1^2\,x_2^2 + 6x_1^2x_4 + 6x_2^3 + 8x_3^2).$$

4 Show that the cycle index of the symmetric group of degree n is equal to the coefficient of z^n in the development of

$$\exp\left(zx_1 + \frac{z^2x_2}{2} + \frac{z^3x_3}{3} + \cdots\right)$$

as a power series in z.

5 a) Let S be any finite group of order n. If a is a fixed element of S, show that the mapping $s \to as$ is a permutation of S.

b) Denoting this permutation by g_a, prove that if a runs through S the g_a's form a permutation group G. Show that $g_a g_b = g_{ab}$. The group G is called the *Cayley representation* of the group S.

c) Show that the cycle index

$$P_G = \frac{1}{m} \sum_{a \text{ in } S} (x_{k(a)})^{n/k(a)},$$

where $k(a)$ is the order of the element a.

d) Show that the cycle index can also be written as

$$P_G = \frac{1}{n} \sum_{d/n} \nu(d)\, x_d^{n/d},$$

where d runs through the divisors of n, and (d) represents the number of elements a in S with order $k(a) = d$.

6 a) Let G be the group of rotations of a pyramid (tetrahedron) around a vertical axis from the apex to the center of the base which leave it fixed. Then G is of order 3 and is generated by a counterclockwise 120° rotation. The group G induces a group of permutations acting on the faces, also denoted by G. Show that

$$P_G = \tfrac{1}{3}(x_1^4 + 2x_1x_3).$$

b) Show that the group G of rotations which leave a regular tetrahedron fixed consists of:

i) the identity permutation.
ii) 8 permutations corresponding to 120° rotations around lines connecting a vertex and the center of its opposite face.
iii) 3 permutations corresponding to 180° rotations around lines connecting the midpoints of opposite edges.

c) If G is considered acting on the vertices of the tetrahedron show that

$$P_G = \tfrac{1}{12}(x_1^4 + 8x_1x_3 + 3x_2^2).$$

6 Polya's theorem

The direct application of Burnside's lemma is sometimes impractical for a complex permutation group or a large set. This is due to the fact that the computation of the number of invariances under the permutations can be both difficult and tedious. Moreover, it is frequently necessary to have more information than merely the number of equivalence classes. For example, if each corner of a cube is painted red or black it might be required to find the number of cubes with 6 black corners and 2 red corners (cf. exercise 4.3). The first difficulty is overcome by the use of the cycle index; the second by assigning weights. It will be seen that if the weights b or r are assigned to a corner if it is painted black or red respectively then Burnside's lemma can be modified to yield a generating function. The coefficient of b^6r^2 in the generating function is the number of cubes with 6 black corners and 2 red corners. This elegant merging of the notions of equivalence introduced by groups, weights and generating functions is due to Polya.

To formulate and prove Polya's theorem in an abstract and concise manner it is convenient to regard the objects to be counted as the set of all mappings of a finite set D into a finite set R.

Example 1

1 The selection of any number of integers from the set $D = \{1, 2, 3, 4, 5\}$ can be regarded as a mapping of D into $R = \{0, 1\}$ (cf. example 1.7.1).

2 A 3 permutation (allowing repetitions) which can be formed from the letters a, b, c can be regarded as a mapping of the set $D = \{a, b, c\}$ into $R = \{a, b, c\}$—e.g., (a, a, a) denotes the mapping $f(a) = f(b) = f(c) = a$.

3 Let D consist of all 8 corners of a cube and R consist of the two words red and black. Each function on D into R can be regarded as a way of painting the cube so that each corner is painted either red or black.

In general, consider the set of all functions defined on a finite set D with values in a finite set R. The set of all such functions will be denoted R^D. The range R is sometimes called the *store*. Each function f in R^D is constructed by assigning to each d in D an element $f(d)$ in R. Therefore, it can be regarded as an r-permutation ($r = |D|$ and $n = |R|$) with repetitions. Hence the number of elements of R^D is $|R|^{|D|}$ by theorem 1.3.3.

As in the last section, we will introduce a group of permutations of R^D. This is accomplished by a group G of permutations on D. Every g in G defines a mapping π_g of R^D into R^D as follows:

$$\pi_g(f) = fg^{-1}. \tag{1}$$

It is obvious that for a fixed g, π_g is a 1–1 mapping of R^D onto itself and so by definition is a permutation. To show that these permutations form a group note that

$$\pi_{gh} = \pi_g \pi_h \tag{2}$$

since for every f in R^D,

$$\pi_g \pi_h (f) = \pi_g (fh^{-1}) = fh^{-1}g^{-1} = f(gh)^{-1} = \pi_{gh}(f).$$

Hence the closure law holds. If e is the identity for the group G, then $\pi_e f = f$. It follows from (2) that the inverse law is valid—in fact, $\pi_g^{-1} = \pi_{g^{-1}}$. Since the associative law is always valid for mappings (c.f. example 3.5) the permutations form a group.

As in the last section, the group of permutations decompose the set of functions into nonoverlapping subsets which are the equivalence classes (cf. theorem 4.1). Two functions are equivalent (i.e. in the same equivalence class) if and only if there is a permutation which carries one into the other. The equivalence classes are called *patterns*.

Example 2 Consider example 1, part 3. Let G be the group of all permutations of D that can be produced by rotations of the cube (see example 5.2). If one of two differently painted cubes can be rotated in such a manner that they appear to be the same, then the two cubes belong to the same pattern.

The 23 patterns are listed below by drawing a typical cube in a pattern (see Fig. on p. 202). The number of cubes in each pattern is given in parenthesis. (p), (q), (r), (s), (t), (u), (v) and (w) obtained from (a)–(h) respectively on interchanging the words red and black.

The concept of weight will be introduced next. Each element r in R is assigned a *weight* $W(r)$ which is a number or a variable. Sums, products and rational multiples of weights can be formed and these operations satisfy the usual associative, commutative and distributive laws. Hence the following definition is sensible.

DEFINITION 1 *The* weight of a function *f in R^D, denoted by $W(f)$, is defined by*

$$W(f) = \prod_{d \text{ in } D} W(f(d)).$$

As an example, let $D = \{d_1, d_2\}$, $R = \{r_1, r_2\}$, $W(r_1) = a$ and $W(r_2) = b$. The weight of the function f defined by $f(d_1) = f(d_2) = r_1$ is a^2. The weight of the function g defined by $g(d_1) = r_1$ and $g(d_2) = r_2$ is ab.

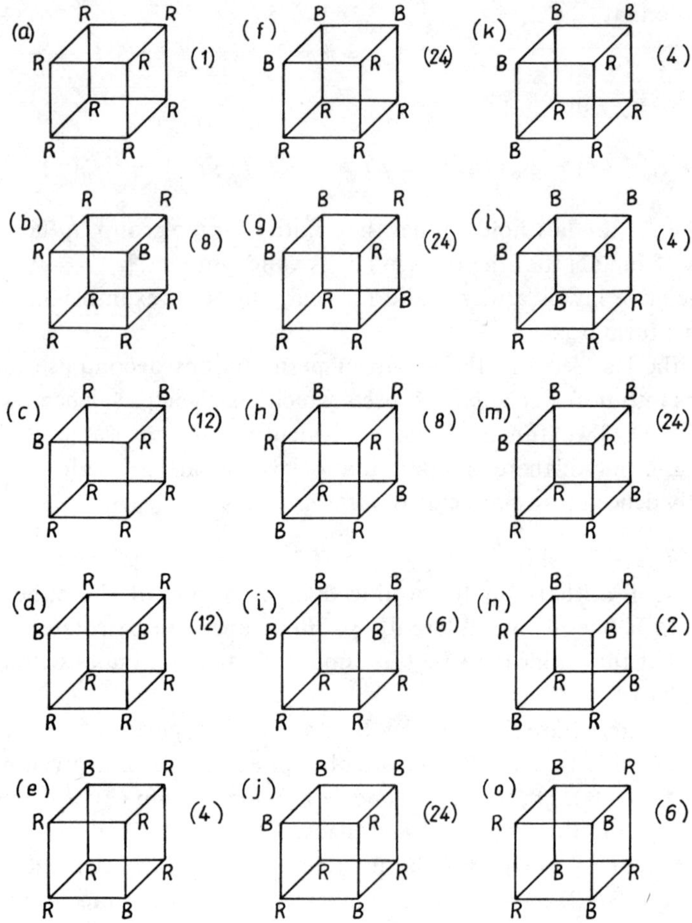

Lemma 1 *If f_1 and f_2 are equivalent (i.e. they belong to the same pattern), then*

$$W(f_1) = W(f_2).$$

Proof By hypothesis, $f_1 g = f_2$ for some g in G. Therefore

$$W(f_1) = \prod_{d \text{ in } D} W(f_1(d)) = \prod_{d \text{ in } D} W(f_1(g(d))) = \prod_{d \text{ in } D} W(f_2(d)) = W(f_2),$$

since the first and second product have the same factors and the order of the factors is immaterial.

This lemma justifies the following definition.

DEFINITION 2 *The* weight of a pattern P, *denoted by* $W(P)$, *is defined by* $W(P) = W(f)$ *where* f *is a (fixed) function in* P.

Example 3

1 Let $W(r) = 1$ for all r in R. Then $W(f) = 1$ for all functions and $W(P) = 1$ for all patterns.

2 In example 2, let W (red) $= r$ and W (black) $= b$. Then the 23 patterns (a)–(w) have weights r^8, r^7b, r^6b^2, r^6b^2, r^6b^2, r^5b^3, r^5b^3, r^5b^3, r^4b^4, r^4b^4, r^4b^4, r^4b^4, r^4b^4, r^4b^4, r^4b^4, b^8, rb^7, r^2b^6, r^2b^6, r^2b^6, r^3b^5, r^3b^5, r^3b^5, respectively. Note that different patterns need not have different weights.

DEFINITION 3 *Let S be an arbitrary set whose elements s have weight $W(s)$. The* inventory of S, *denoted by* inv (S), *is defined by*

$$\text{inv}\,(S) = \sum_{s \text{ in } S} W(s).$$

Example 4

1 In example 3, part 1, inv $(R) = \sum_{r \text{ in } R} W(r) =$ number of elements in R and inv $(R^D) = \sum_{f \text{ in } R^D} W(f) =$ number of functions in R^D. Since the number of functions in R^D is $|R|^{|D|}$ it follows that inv $(R^D) = (\text{inv}\,(R))^D$.

2 In example 3, part 2,

$$\text{inv}\,(R^D) = r^8 + 8r^7b + 28r^6b^2 + 56r^5b^3 + 70r^4b^4 + 56r^3b^5$$
$$+ 28r^2b^6 + 8rb^7 + b^8 = (r + b)^8.$$

These examples lead to the following conjecture.

LEMMA 2

$$\text{inv}\,(R^D) = (\text{inv}\,R)^{|D|}$$

Proof By definition, $(\text{inv}\,R)^D = \left(\sum_{r \text{ in } R} W(r)\right)^{|D|}$ which can be written as the product of $|D|$ identical factors. A term of the expansion of this product is obtained by selecting one term from each factor and multiplying them together. The full product expansion is obtained by adding all such terms. Suppose D contains the points $d_1, d_2, \ldots, d_n$. Define a function f in R^D as follows: if $W(r)$ is selected from the ith factor, let $f(d_i) = r$. Hence to f there corresponds the term $\sum_{d \text{ in } D} W(f(d)) = W(f)$ of the full product expansion.

Since, on the other hand, each f in R^D determines a choice of terms from the factors, the full product expansion equals $\sum_{f \text{ in } R^D} W(f) = \text{inv}(R^D)$.

To prove Polya's theorem, a generalization of the above lemma is required.

LEMMA 3 *Suppose that a finite set D is decomposed into k nonoverlapping subsets $D_1, D_2, \ldots, D_k$. Let S be the set of all functions on D into a finite set R which are constant on each subset D_i $(i = 1, 2, \ldots, k)$. Then*

$$\text{inv}(S) = \prod_{i=1}^{k} \sum_{r \text{ in } R} [W(r)]^{|D_i|}. \tag{3}$$

Proof If f is in S, then $f(d) = c_i$ for all d in D_i. Let $\phi(i) = c_i$ $(i = 1, 2, \ldots, k)$ and let $\psi(d) = i$ for all d in D_i; $f = \phi\psi$.

The remaining part of the proof is similar to the proof of lemma 1; first expand the product in (3). A term of the expansion is obtained by selecting a term in each factor of the right side of (1) which is equivalent to selecting a mapping of the set $\{1, \ldots, k\}$ into R. This produces the term

$$[W(\phi(1))]^{|D_1|} \cdots [W(\phi(k))]^{|D_k|} = \prod_{i=1}^{k} [W(\phi(i))]^{|D_i|}.$$

If $f = \phi\psi$, the fact that $[W(\phi(i))]^{|D_i|} = \sum_{d \text{ in } D_i} W(\phi(\psi(d)))$ implies that

$$\prod_{i=1}^{k} [W(\phi(i))]^{|D_i|} = \prod_{i=1}^{k} \sum_{d \text{ in } D_i} W(f(d)) = W(f).$$

In this way, each f in S is obtained exactly once and so the full expansion of the product is the sum of all such terms or $\sum_{f \text{ in } S} W(f) = \text{inv}(S)$.

Example 5 Suppose that a blue, red, yellow and green die are cast. Find the number of ways that the resulting total is 14 and the blue and red die have the same number of spots on their uppermost face.

Let $D = \{b, r, g, y\}$ and $R = \{1, 2, 3, 4, 5, 6\}$. A function f on D into R can obviously be interpreted as giving the number of spots which turn up on a die—for example, $f(b) = 3, f(r) = 2, f(g) = 5, f(y) = 6$ denotes the fact that the blue, red, green and yellow die have 3, 2, 5, and 6 spots respectively on their uppermost face. The solution of the problem is the number of functions such that $f(r) = f(b)$ and

$$f(r) + f(b) + f(g) + f(y) = 14. \tag{4}$$

We shall use lemma 3 to find this number. Let S denote the set of functions which are constant on $D_1 = \{b, r\}$, $D_2 = \{y\}$, and $D_3 = \{g\}$. Assign the weights $x, x^2, \ldots, x^6$ to the elements $1, 2, \ldots, 6$ of R respectively. Then by (3), the inventory of all functions that are constant on each D_i is

$$\text{inv}(S) = (x^2 + x^4 + x^6 + x^8 + x^{10} + x^{12})(x + x^2 + x^3 + x^4 + x^5 + x^6)^2.$$

The required number is the coefficient of x^{14} in this expansion, since the functions which satisfy (4) have weight x^{14}.

LEMMA 4 *Let a finite set D be split into cycles by a permutation g. Then a function f on D is constant on every cycle of g if and only if $f = fg$.*

Proof Suppose $f = fg$. A cycle of D is of the form $d, g(d), \ldots, g^k(d)$. By hypothesis, $f(d) = f(g(d))$; similarly, $f(g(d)) = fg(g(d)) = f(g^2(d))$. Hence $f(d) = f(g^2(d))$. Repeating this procedure yields the result that f is constant on the cycle.

Conversely, suppose that f is constant on every cycle. Since every d in D belongs to some cycle and $g(d)$ belongs to the same cycle, $f(d) = f(g(d))$ for any d in D.

Before proving Polya's theorem, we briefly summarize some of the concepts involved. Let D and R be finite sets and G be a group of permutations on D. The set R^D of all functions on D into R is divided into equivalence classes, called patterns, by placing two functions in the same pattern if and only if $f_1 = f_2 g$ for some g in G. If each element r of R is assigned a weight $W(r)$ then the weight $W(f)$ of a function f is defined by definition 1 and the weight $W(P)$ of a pattern P by definition 2. The inventory $\Sigma\, W(P)$ of the set of all patterns is given by Polya's Fundamental Theorem.

THEOREM 1 *The pattern inventory is*

$$\sum_P W(P) = P_G\left(\sum_{r \text{ in } R} W(r), \sum_{r \text{ in } R} [W(r)]^2, \sum_{r \text{ in } R} [W(r)]^3, \ldots\right), \tag{5}$$

where P_G is the cycle index.

Proof Let S be the set of all functions f in R^D satisfying $W(f) = w$ where w is one of the possible values that the weight of a function may have. Decompose S into equivalence classes by means of the permutations π_g (see (1)) for g in G. By Burnside's lemma (theorem 4.2) the number of equivalence classes in S (or the number of patterns as far as they are contained in S) equals

$$|G|^{-1} \sum_{g \text{ in } G} I_w(g) \tag{6}$$

where $I_w(g)$ denotes the number of functions, with $W(f) = w$, left fixed by π_g (that is, $f = fg$). The patterns contained in S all have weight w. Therefore, multiplying (6) by w and summing over all possible values of w yields the pattern inventory

$$\sum W(P) = |G|^{-1} \sum_{w} \sum_{g \text{ in } G} I_w(g)\, w = |G|^{-1} \sum_{g \text{ in } G} \sum_{w} I_w(g)\, w .$$

Since

$$\sum_{w} I_w(g)\, w = \sum_{(f:f=fg)} W(f), \tag{7}$$

where the summation is over all f in R^D which satisfy $f = fg$, it follows that

$$\sum W(P) = |G|^{-1} \sum_{g \text{ in } G} \sum_{(f:f=fg)} W(f). \tag{8}$$

Let $D_1, \ldots, D_k$ be the sets containing the elements of the cycles of g. By lemma 4, f is constant on every cycle of g. Hence the sum in the right side of (7) is just the inventory of the set of all functions constant on each D_i ($i = 1, 2, \ldots, k$). This inventory is given by (3). To evaluate this expression, let $\{k_1, k_2, \ldots\}$ be the type of g. Then among the numbers $|D_1|, \ldots, |D_k|$, the number 1 occurs k_1 times, the number 2 occurs k_2 times, etc. Therefore the sum in the right side of (7) equals

$$\Big[\sum_{r \text{ in } R} W(r)\Big]^{k_1} \Big[\sum_{r \text{ in } R} (W(r))^2\Big]^{k_2} \Big[\sum_{r \text{ in } R} (W(r))^3\Big]^{k_3} \cdots .$$

The above expression can be obtained by substitution of

$$x_1 = \sum_{r \text{ in } R} W(r), \quad x_2 = \sum_{r \text{ in } R} (W(r))^2, \quad x_3 = \sum_{r \text{ in } R} (W(r))^3, \ldots$$

into the product $x_1^{k_1} x_2^{k_2} x_3^{k_3} \cdots$, which is the term corresponding to g in $|G|\, P_G$ (see definition 5.3). Hence, by definition 5.3, the right side of (8) is just $P_G\,(x_1, x_2, x_3, \ldots)$; this proves Polya's theorem.

COROLLARY *The number of patterns equals* $P_G\,(|R|, |R|, |R|, \ldots)$.

The corollary follows immediately from the theorem by choosing all the weights equal to one.

Example 6

1 By the corollary to Polya's theorem, the number of color schemes in example 2 is $C_G\,(2, 2, 2, \ldots)$. By example 5.2

$$P_G\,(2, 2, 2, \ldots) = \tfrac{1}{24}\,(2^8 + 9 \cdot 2^4 + 6 \cdot 2^2 + 8 \cdot 2^4) = 23 .$$

The different patterns can be enumerated by means of the pattern inventory. Polya's theorem with $W(\text{red}) = r$ and $W(\text{blue}) = b$ yields

$$\tfrac{1}{24}\left[(r + b)^8 + 9(r^2 + b^2)^4 + 6(r^4 + b^4)^2 + 8(r + b)^2(r^3 + b^3)^2\right].$$

2 Let D be a finite set, and let G be a permutation group of D. Two subsets D_1 and D_2 are equivalent if for some g in G the set of elements $g(d_1)$ as d_1 runs through D_1 is equal to D_2. As usual, equivalence classes of subsets can be formed. We will find the number of such classes. Subsets of D can be brought into 1–1 correspondence with functions as follows. Let $R = \{0, 1\}$ and assign the weight 1 to each. If f is in R^D, then to f corresponds the subset of all d in D with $f(d) = 1$. All subsets are obtained in this way. Moreover, functions are equivalent if and only if corresponding subsets are equivalent. Since each function has weight 1, the number of equivalence classes of subsets equals the pattern inventory, $P_G(2, 2, \ldots)$.

Burnside's lemma (theorem 4.2) can be generalized by taking $W(0) = 1$ and $W(1) = x$. Then the subsets containing k elements correspond to a function f with $W(f) = x^k$. Hence the number of classes consisting of subsets with k elements each equals the coefficient of x^k in the pattern inventory, $P_G(1 + x, 1 + x^2, 1 + x^3, \ldots)$. In particular, for $k = 1$, the number of equivalence classes in the sense of Burnside's lemma is obtained.

Example 7 Consider an alphabet of k different letters. Ordered selections of n letters (words) formed from this alphabet can be regarded as functions whose domain $D = \{1, 2, \ldots, n\}$ and whose range is the alphabet. Let the weight of the ith letter be x_i $(i = 1, 2, \ldots, k)$. Ordered choices of these letters may be considered in various ways.

1 The simplest situation is that in which all orders are regarded as different—in other words, the group of permutations consists of the identity. Since its cycle index is t_1^n, the pattern inventory is

$$(x_1 + x_2 + \cdots + x_k)^n = \frac{n!}{i_1!\, i_2! \cdots i_k!}\, x_1^{i_1} x_2^{i_2} \cdots x_k^{i_k}$$

with $i_1 + i_2 + \cdots + i_k = n$. The coefficient $n!/i_1!\, i_2! \cdots i_k!$ is the number of permutations of n letters, i_1 of which are of one kind, i_2 of a second kind and so on (cf. theorem 1.3.2).

2 The problem of finding the number of necklaces with n beads where i_j beads are all of color j $(j = 1, 2, \ldots, k)$ is equivalent to placing n letters on n equidistant points of a cirele where i_j letters are of the jth kind. We consider all possible ordered selections of n letters from an alphabet of k different letters. Two selections are regarded as equivalent if they can be made to coincide under a rotation of the circle. The group is the cyclic group with cycle index $C_n(t_1, t_2, \dots, t_n)$. (See example 5.4.) Hence the pattern inventory is

$$C_n(s_1, \dots, s_n) = n^{-1} \sum_{d|n} \phi(d)\, s_d^{n/d}$$

where $s_r = x_1^r + x_2^r + \cdots + x_k^r$. Expanding $s_d^{n/d}$ by the multinomial theorem yields

$$C_n(s_1, \dots, s_n) = \sum K(i_1, \dots, i_k)\, x_1^{i_1} x_2^{i_2} \cdots x_k^{i_k}$$

where

$$K(i_1, \dots, i_k) = \frac{1}{n} \sum \phi(d) \frac{(n/d)!}{(i_1/d)! \cdots (i_k/d)!} \tag{9}$$

and the sum is over all divisors of the greatest common divisor of the nonzero members of the set i_1 to i_k. $K(i_1, \dots. i_k)$ is the number of necklaces with i_j beads of the jth color.

Exercises

1 Solve the following problems by using the notion of the inventory of a set of functions.

a) Find all possible ways of painting three chairs in solid colors with red, white or blue paint.

b) Two families, each having two members, and three other people plan to visit one of three cities. If the members of a family must travel together, find the number of ways the seven people can plan their trips.

2 A string of length three is composed of red and blue beads. Two strings are indistinguishable if interchanging the ends of one will yield the other.

a) Show how to interpret a string of beads as a function from $D = \{1, 2, 3\}$ to $R = \{r, b\}$.

b) Let $W(b) = b$ and $W(r) = r$. Show that the pattern inventory is

$$b^3 + 2b^2r + 2br^2 + r^3.$$

3 Show that the pattern inventory for the ways of painting the four faces of a pyramid (see exercise 5.6 part (a)) with two colors r and b is

$$r^4 + b^4 + 2r^3b + 2r^2b^2 + 2rb^3.$$

4 Consider the class of organic molecules of the form

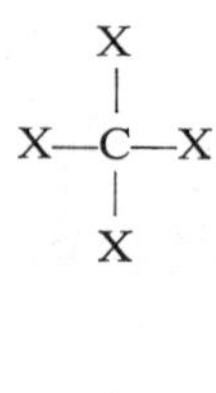

where C is a carbon atom and each X denotes any one of the components chlorine (Cl), ethyl (C_2H_5), hydrogen (H) or methyl (CH_3). Each such molecule can be represented as a regular tetrahedron with the carbon atom at the center and the components labeled X at the corners. Two molecules are equivalent if one can be rotated into the other. (See exercise 5.6 part (b)).)

a) Find the number of different molecules.

b) Show that the number of molecules containing one or more hydrogen atoms is 21.

c) Show that there is one molecule containing four hydrogen atoms, three molecules containing three hydrogen atoms, six molecules containing two hydrogen atoms, 11 molecules containing one hydrogen atom and 15 molecules containing no hydrogen atoms.

5 How many different patterns are there if each face of a cube is painted either red or blue (see exercise 5.3)?

6 In how many ways can a cube be colored so that it has 4 red and 2 blue faces?

7 Deduce the result derived in example 4.2 from part 2 of example 7.

8 Use exercise 5.3 and theorem 2 to deduce the generating function for combinations with repetitions. [Hint: $\Sigma\, T_n(x_1, \ldots, x_k)\, t^n = \exp(s_1 t + s_2 t^2/2! + \cdots + s_n t^n/n! + \cdots)$ $= \exp \sum_{i=1}^{k} (x_i t + x_i^2 t^2/2 + \cdots) = \exp \log (1 - x_1 t)^{-1} \cdots (1 - x_k t)^{-1}$.]

9 Suppose that letters are arranged on a circle (as in example 7 part 2) but equivalence is defined by using rotations and a reflection (the group is now the dihedral group). The distinct arrangements are called bicircular words. Find the total number of bicircular words of n letters for a k-letter alphabet.

7 Generalization of Polya's theorem

In the last section, mappings of D into R with an equivalence introduced by a permutation group G of D were considered. Here a more general situation in which there is also a permutation group H of R is discussed. Every g in G and h in H defines a mapping π_{gxh} of R^D into R^D as follows:

$$\pi_{gxh}(f) = hfg^{-1}. \tag{1}$$

It is obvious that for fixed g and h, π_{gxh} is a 1–1 mapping of R^D onto itself and so by definition is a permutation. To show that these permutations form a group note that

$$\pi_{g_1xh_1}\pi_{g_2xh_2} = \pi_{g_1g_2xh_1h_2}, \tag{2}$$

since for every f in R^D

$$\pi_{g_1xh_1}\pi_{g_2xh_2}(f) = \pi_{g_1xh_1}(h_2fg_2^{-1}) = h_1h_2fg_2^{-1}g_1^{-1} = (h_1h_2)f(g_1g_2)^{-1}$$
$$= \pi_{g_1g_2xh_1h_2}(f).$$

Hence the closure law holds. If 1 and e are the identities for G and H respectively, then $\pi_{1xe}(f) = f$. It follows from (2) that the inverse law is valid—in fact,

$$\overset{-1}{\pi_{gxh}} = \pi_{g^{-1}x^{-1}}.$$

Since the associative law is always valid for mappings (cf. example 3.5) the permutations form a group.

This group of permutations decompose the set of functions into nonoverlapping subsets which are the equivalence classes or patterns. Again, two functions are said to be equivalent if and only if they belong to the same pattern.

In general, equivalent functions do not have the same weight; this is shown by the following example. Let $D = \{d_1, d_2\}$ and $R = \{r_1, r_2\}$. The permutation group G of the domain D contains the permutations $g_1 = e$ and $g_2 = (d_1, d_2)$ and the permutation group H of the range R contains the permutation $h_1 = e$ and $h_2 = (r_1, r_2)$. The function f_1 defined by $f_1(d_1) = f_1(d_2) = r_1$ and the function f_2 defined by $f_2(d_1) = f_2(d_2) = r_2$, are equivalent since $h_2 f_1 = f_2 g_1$. However, if $W(r_1) = r_1$ and $W(r_2) = r_2$ then $W(f_1) = r_1^2$ while $W(f_2) = r_2^2$.

In order to talk about the weight of a pattern and the pattern inventory, an additional condition must be imposed. Weights should be assigned to the elements in the range R in such a way that functions in the same equivalence class will have the same weight. There is a fairly general but complicated way of doing this. To simplify the exposition, the weight 1 will be assigned to each element r in R; then $W(f) = 1$ for each f in R^D. The condition that weights of functions in the same equivalence class are the same is therefore trivially satisfied. It follows that the pattern inventory will then be the number of equivalence classes.

The number of equivalence classes can easily be obtained from Burnside's lemma. The order of the group of elements $\{\pi_{gxh}\}$ with g in G and h in H is $|G|\ |H|$ since each element π_{gxh} corresponds uniquely to an element gxh of GxH, the product of G and H (see example 5.6) and conversely. Therefore, Burnside's lemma yields the following result.

THEOREM 1 *The number of equivalence classes of functions from D to R is*

$$|G|^{-1}\ |H|^{-1} \sum_{g \text{ in } G} \sum_{h \text{ in } H} I(\pi_{gxh}) \tag{3}$$

where $I(\pi_{gxh})$ is the number of functions f satisfying

$$hf(d) = f(g(d))$$

for all d in D.

To simplify (3), $I(\pi_{gxh})$ must be evaluated. This requires the following lemma.

LEMMA 1 *f in R^D is invariant* under π_{gxh} if and only if f maps every cycle*

$$(d, g(d), \ldots, g^{j-1}(d)) \tag{4}$$

of elements of D onto a cycle

$$(f(d), h(f(d)), \ldots, h^{i-1}(f(d)) \tag{5}$$

of elements of R where i is a divisor of j.

Proof A function which satisfies the hypothesis of the lemma is clearly invariant under π_{gxh}.

Conversely, assume that f is invariant under π_{gxh}; that is $fg = hf$. This implies that $fg^2 = fgg = hfg = hhf = h^2f = fg^3 = h^3f$, and so on. Hence f maps the elements in (4) onto those in (5). Since $h^j(f(d)) = f(g^j(d)) = f(d)$ it follows that the length i of the cycle of R to which $f(d)$ belongs is a divisor of j.

THEOREM 2 *The number of patterns of mappings of D into R (patterns being defined by the equivalence induced by permutation groups G of D and H of R according to (1)) equals*

$$P_G\left(\frac{\partial}{\partial z_1}, \frac{\partial}{\partial z_2}, \frac{\partial}{\partial z_3}, \cdots\right) P_H\left(e^{z_1+z_2+z_3+\cdots}, e^{2(z_2+z_4+z_6+\cdots)}, e^{3(z_3+z_6+z_9+\cdots)}, \ldots\right)$$

evaluated at $z_1 = z_2 = z_3 = \cdots = 0$.

Proof As noted after theorem 1, $I(\pi_{gxh})$ must be evaluated. Select an element d in a cycle of length j of g which generates the cycle, that is all elements of the cycle appear in (4). According to lemma 1, this cycle is mapped onto a cycle of length i in h with i being a divisor of j; moreover, according to (5), once the image of d is specified the images of the other elements in the cycle are completely determined. Hence it is only necessary to find the possible images of d to determine functions f which are invariant under π_{gxh}. Since there are i possible images for d in a cycle of length i and there are c_i cycles of length i, where $\{c_1, c_2, \ldots\}$ is the type of h, the total number of possibilities for d is

$$\sum_{i|j} ic_i \tag{6}$$

---

* $\pi_{gxh}(f) = f$.

where the sum is taken over all divisors i of j. Let $\{b_1, b_2, \ldots\}$ be the type of g; there are b_j cycles of length j. Since the choice of functional values for each selected element in a cycle of g is independent and because these choices determine f completely, the number of functions which are invariant under π_{gxh} equals the product of (6) taken over all selected elements. Hence

$$I(\pi_{gxh}) = \prod_j \Big(\sum_{i|j} ic_i\Big)^{b_i} = (c_1)^{b_1}(c_1 + 2c_2)^{b_2}(c_1 + 3c_3)^{b_3}(c_1 + 2c_2 + 4c_4)^{b_4}\cdots$$

Since

$$(c_1)^{b_1} = \left(\frac{\partial}{\partial z_1}\right)^{b_1} e^{c_1z_1}|_{z_1=0}$$

$$(c_1 + 2c_2)^{b_2} = \left(\frac{\partial}{\partial z_2}\right)^{b_2} e^{c_1z_2}e^{2c_2z_2}|_{z_2=0}$$

$$(c_1 + 3c_3)^{b_3} = \left(\frac{\partial}{\partial z_3}\right)^{b_3} e^{c_1z_3}e^{3c_3z_3}|_{z_3=0}$$

$$(c_1 + 2c_2 + 4c_4)^{b_4} = \left(\frac{\partial}{\partial z_4}\right)^{b_4} e^{c_1z_4}e^{2c_2z_4}e^{4c_4z_4}|_{z_4=0}$$

. . . .

it follows that

$$I(\pi_{gxh}) = \left[\left(\frac{\partial}{\partial z_1}\right)^{b_1}\left(\frac{\partial}{\partial z_2}\right)^{b_2}\cdots\left(\frac{\partial}{\partial z_k}\right)^{b_k}\cdots\right] \times$$
$$\times\ [e^{c_1(z_1+z_2+z_3+\cdots)}\, e^{2c_2(z_2+z_4+z_6+\cdots)}\cdots$$
$$\cdots e^{kc_k(z_k+z_{2k}+z_{3k}+\cdots)}\cdots]$$

evaluated at $z_1 = z_2 = \cdots = z_k = 0$. The theorem is now an immediate consequence of (3).

Example 1 In how many ways can five gifts, two of which are the same, be placed in two identical boxes and two other dissimilar boxes?

Let $D = \{g_1, g_2, g_3, g_4, g_5\}$ denote the set of five gifts with g_1 and g_2 representing the gifts which are alike. Since two ways of distributing the gifts should be considered equivalent whenever g_1 and g_2 are interchanged, let $G = \{e, (g_1, g_2)\}$ be the permutation group of D. Let $R = \{b_1, b_2, b_3, b_4\}$ represent the four boxes with b_1 and b_2 being the similar boxes. Since two ways of placing the gifts are equivalent if the gifts in the identical boxes are interchanged, let $H = \{e, (b_1, b_2)\}$ be the permutation group of R. The number of patterns is

$$\frac{1}{4}\left[\frac{\partial^5}{\partial z_1^5} + \frac{\partial^3}{\partial z_1^3}\frac{\partial}{\partial z_2}\right][e^{4(z_1+z_2)} + e^{2(z_1+z_2)}e^{2z_2}]|_{z_1=z_2=0} = 336.$$

Exercises

1 Let $D = \{a, b, c\}$, $G = S_3$, $R = \{r, s\}$ and $H = \{e, (r, s)\}$. Show that there are two patterns.

2 Each square of a 2×2 chessboard is painted either black or white. If one chessboard cannot be distinguished from a second after a rotation through 90°, 180° or 270° the two are equivalent. Two chessboards are also to be regarded as equivalent if one can be obtained from the other by replacing each square by a square of the opposite color. Find the number of nonequivalent chessboards.

3 Suppose that every vertex of a tetrahedron is connected by a wire to a vertex of a cube. Two different wires can lead to one and the same vertex of the cube but just one wire is attached to each vertex of the tetrahedron. Two possibilities are called equivalent if the first one can be transformed into the other one by rotations of the solids. Only connections are taken into consideration and not the way the wires are interlinked in the space between the solids. Show that there are 27 patterns.

4 A message consists of a sequence of n digits where each digit is either 1, 2, 3, or 4. Each digit of a sequence is indicated by the flashing of one of four lights.

a) Suppose the labels on lights 1 and 2 have been removed. Show that there are $\frac{1}{2}(4^n + 2^n)$ distinguishable messages.

b) In addition, suppose that occasionally the first two digits of a sequence are interchanged before transmission. Show that there are now $\frac{1}{4}(4^n + 4^{n-1} + 2^{n+1})$ distinguishable messages.

5 Let G be a permutation group of D and H a permutation group of R.

a) Let g be in G, h be in H, and f be a function in R^D. Show that if

$$f(g(d)) = hf(d) \quad \text{(for all } d \text{ in } D)$$

then

$$f(g^j(d)) = h^j f(d) \quad \text{(for all } d \text{ in } D;\ j = 1, 2, \ldots)$$

b) If f is $1-1$ and $f(g(d)) = hf(d)$ for all d in D, prove that f maps any element in D that belongs to a cycle of length j in g onto an element in R that belongs to a cycle of length j in h.

c) Let g be of type $\{b_1, b_2, \ldots\}$ and h be of type $\{c_1, c_2, \ldots\}$. Show that the number of $1-1$ functions f which are such that $f(g(d)) = hf(d)$ for all d in D is equal to $\prod_j j^{b_j} c_j!/(c_j - b_j)!$ if $c_j \geq b_j$ and is zero otherwise. Then show that

$$\prod_i j^{b_j} c_j!/(c_j - b_j)! = \left(\frac{\partial}{\partial z_1}\right)^{b_1}\left(\frac{\partial}{\partial z_2}\right)^{b_2}\cdots(1 + z_1)^{c_1}(1 + 2z_2)^{c_2}\cdots$$

evaluated at $z_1 = z_2 = \cdots = 0$.

d) Show that the number of patterns of $1-1$ functions from D to R, with equivalence induced by permutation groups G on D and H on R, is equal to

$$P_G\left(\frac{\partial}{\partial z_1},\frac{\partial}{\partial z_2},\cdots\right)P_H(1+z_1,1+2z_2,1+3z_3,\ldots)$$

evaluated at $z_1 = z_2 = \cdots = 0$.

e) Show that if $|R| = |D|$, (d) reduces to

$$P_G\left(\frac{\partial}{\partial z_1},\frac{\partial}{\partial z_2},\ldots\right)P_G(z_1,2z_2,3z_3,\ldots).$$

References

Möbius Function

1 Weisner, L., "Abstract theory of inversion of finite series", *Trans. Amer. Math. Soc.*, **38** (1935).

2 Hall, P., "The Eulerian function of a group", *Quart. J. Math. Ox.*, Series 7 (1936).

3 Rota, G.C., "On the foundations of combinatorial theory, I". Theory of Möbius functions, Z. *Wahrscheinlichkeitstheorie und verw. Gebiete*, **2** (1964).

4 Crapo, H.H., "The Möbius function of a lattice", *J. Comb. Theory* **1** (1966).

Burnside's lemma and applications

1 Burnside, W., *Theory of Groups of Finite Order*, 2nd edition, Cambridge Univ. Press, Cambridge, England (1911).

2 Redfield, J.H., "The theory of group reduced distributions", *Am. J. Math.* **49** (1927).

3 Golomb, S.W., "A mathematical theory of discrete classification", *Information Theory, 4th Lond. Symp.* Cherry, London, Butterworths (1961).

Polya's theorem

1 Polya, G., "Kombinatorische Anzahlbestimmungen für Gruppen, Graphen, und Chemische Verbindungen", *Acta Math.*, **68** (1937).

Applications of Polya's theorem

1 Gilbert, E.N., and Riordan, J., "Symmetry types of periodic sequences", *Ill. J. Math.*, vol. 5 (1961).

2 Riordan, J., "The combinatorial significance of a theorem of Polya", *J. Soc. Ind. Appl. Math.*, vol. 5, No. 4 (1957).

3 Harary, F., "The Number of linear, directed, rooted and connected graphs", *Trans. Am. Math. Soc.*, **78**, pp. 445–463 (1955).

4 Harrison, M.A., *Introduction to Switching and Automata Theory*, McGraw-Hill Book Co., New York (1965) (see chapter 5).
5 Polya, G., "Sur les Types des Propositions Composees", *J. Symbolic Logic*, **5** pp.98–103 (1940).
6 Slepian, D., "On the number of symmetry types of Boolean functions of *n* variables", *Can. J. Math.*, **5** pp.185–193 (1953).

Generalizations of Polya's theorem

1 de Bruijn, N.G., "Generalization of Polya's fundamental theorem in enumerative combinatorial analysis", *Indigationes Mathematicae*, **21** (1959).
2 de Bruijn, N.G., "Enumerative combinatorial problems concerning structures", *Nieuw Archief vor Wiskunde* (**3**), XI (1963).
3 de Bruijn, N.G., "Polya's theory of counting", *Applied Combinatorial Mathematics*, [E.F.Beckenbach editor], John Wiley & Sons, New York (1964).

Bibliography

Combinatorial analysis

1. H. Hadwiger, H. DeBrunner and V. Klee, *Combinatorial Geometry in the Plane*, Holt, Rinehart and Winston, New York (1964).
2. M. Hall, *Combinatorial Theory*, Blaisdell Co., Waltham, Mass. (1967).
3. C. L. Liu, *Introduction to Combinatorial Mathematics*, McGraw-Hill, New York (1968).
4. P. A. MacMahon, *Combinatory Analysis*, Chelsea Publishing Co., New York (1958).
5. J. Niven, *Mathematics of Choice*, Random House Inc., New York (1965).
6. H. J. Ryser, Combinatorial Mathematics, Carus Monograph No. 14, John Wiley, New York (1963).
7. W. A. Whitworth, *Choice and Chance*, Hafner, New York (1959).

   W. A. Whitworth, *DCC Exercises in Choice and Chance*, Hafner, New York (1965).
8. J. Riordan, *An Introduction to Combinatorial Analysis*, John Wiley, New York (1958).

Probability theory

9. M. Eisen, *Introduction to Mathematical Probability Theory*, Prentice-Hall, Englewood Cliffs, New Jersey (1969).
10. W. Feller, *An Introduction to Probability Theory and its Applications*, Vol. 1, 2nd ed., John Wiley, New York (1957).

Answers to problems

Exercise 1.1

a) 1, 2, 3, ..., 10; equally likely; yes
b) all possible hands of 5 cards; equally likely; yes. The events 0, 1, 2, 3, 4 kings are compound events and are not equally likely.
c) 36″, 37″, 38″, ..., 84″; not equally likely; no
d) all infinite sequences of symbols $x_1, x_2, x_3, \ldots$, where $x_i = h$ (heads) or t (tails); equally likely; no. The described event is a compound event.
e) all 10-tuples $(x_1, x_2, x_3, \ldots, x_{10})$ where $x_i = 1$ or 2 or 3 or 4 or 5 or 6 ($i = 1, 2, \ldots, 10$); equally likely; yes. The described event is a compound event.
f) all 2-tuples (a, b) where, $a, b = 1, 2, \ldots, 52$ and $a \neq b$; equally likely; yes.
g) 0, 1, 2, 3, ...; not equally likely; no.
h) 1, 2, 3, ...; equally likely; no.

Exercise 1.2

1 4
3 90
5 a) 2^6; b) 7 if telegrams distributed to boys. Telegrams *delivered* in 7! (or 64 (6!)) ways in (a).
7 $(14)^2 = 196$
9 $24 \cdot 12 \cdot 4 \cdot 6$

Exercise 1.3

1 54
3 May 22, 2064
5 $\dfrac{11!}{4!\,4!\,2!}, \quad \dfrac{10!}{4!\,4!\,2!}, \quad \dfrac{9!}{3!\,4!\,2!}$
7 $m(m-1)$
11 a) $26 \cdot 25 \cdot 24 \cdot 23$ b) $(26)^4$
13 $3^7/52$

Exercise 1.4

1 ${}_{10}C_4$
3 ${}_{10}C_2$ b) $2({}_4C_2)$ c) $\binom{7}{2} + 7 \cdot 4 \cdot 2 + 2\binom{4}{2}$

5 a) $\binom{n}{2}$ b) $\dfrac{n(n-3)}{2}$

7 $\binom{25}{18}\ 2^7$

9 a) $\binom{7}{3}$ b) 5^3

11 a) n^r b) $\binom{n+r-1}{r}$ c) $\binom{n}{r}$

13 a) n^k b) $n(n-1)\cdots(n-k+1)$ c) $\binom{n+k-1}{k}$, $\binom{n}{k}$

15 a) $\binom{15}{8}$ b) ${}_{15}C_7 - {}_8C_4 \cdot {}_6C_3$

19 a) 35 b) 1 c) 21 locks, G gets 20 keys and the others get 11 keys each.

21 $\binom{14}{12}$

Exercise 1.5

1 a) $\binom{64}{61}$ b) $\binom{60}{3}$

3 $1 + \binom{3}{1} + \binom{4}{2} + \binom{5}{3} + \binom{6}{4} + \cdots + \binom{26}{24}$

5 $r = x - 3, s = y - 3, t = z - 3$

7 $\binom{60}{3}$

Exercise 1.6

1 $\dfrac{12!}{(2!)^6\, 6!}$

3 $\dfrac{8!}{(2!)^4\, 4!}$

5

Arrangements	*No. of ways*
(7)	10
(6, 1)	$\dfrac{10!}{8!}\ \dfrac{7!}{6!}$

(5, 2) $\frac{10!}{8!}\ \frac{7!}{5!\,2!}$

(4, 2, 1) $\frac{10!}{7!}\ \frac{7!}{4!\,2!\,1!}$ etc.

Exercises 1.7

1 $2880 = 6 \cdot 5 \cdot 4 \cdot 3 \cdot 2^3$

3 $3 \cdot 5 \cdot 2^2 = 60$

5 8

Exercise 2.1

1 a) $0 + 2 + 6 + 12 + \cdots + n(n - 1)$

b) $n! + (n + 1)! + (n + 2)! + \cdots + (2n)!$

c) $1 + 2x + 3x^2 + \cdots + nx^{n-1}$

d) $1 - 2^3 + 3^3 - 4^3 \cdots - (2n)^3$

e) $2^4 - 1^4 + 3^4 - 2^4 + 4^4 - 3^4 + \cdots + (n + 1)^4 - n^4 = (n + 1)^4 - 1$

f) $1 + \frac{1}{4} + \frac{1}{9} + \cdots + \frac{1}{n^2}$

g) $14 + 22 + 32 + 44 + 58 = 170$

h) $1^{1/3} + 2^{1/3} + 3^{1/3} + \cdots + (m - 1)^{1/3}$

i) $\log 82 - \log s = \log \left(\frac{82}{s}\right)$

3 a) $\sum_{1}^{n} k(k + 2)$

b) $\sum_{1}^{2r} (-1)^{k+1} \frac{1}{k}$

c) $\sum_{r=0}^{k} \frac{1}{(k + r)^3} = \sum_{s=k}^{s=2k} \frac{1}{s^3}$

Exercise 2.2

1 $x^5 + (a + b + c + d + e)x^4 + (ab + ac + ad + ae + bc + bd + be + cd + ce + de)x^3$

$+ (abc + abd + abe + bcd + bce + acd + ace + bde + cde + ade)x^2$

$+ (abcd + bcde + cdea + deab + eabc)x + abcde$

3 a) $\sum_{r=0}^{5} (-1)^r \binom{5}{r} x^{5-r}$

b) $\sum_{r=0}^{3} \binom{3}{r} x^{3-r} (2a)^r$

c) $\sum_{r=0}^{4} \binom{4}{r} x^{2r} (-3y^2)^{4-r}$

d) $\sum_{r=0}^{6} \binom{6}{r} x^{2r-6}$

e) $\sum_{r=0}^{3} \binom{3}{r} (-x)^r y^{6-2r}$

5 There is none.

7 $x^8 - 2x^6 - 2x^4 + 6x^2 - \frac{6}{x^2} + \frac{2}{x^4} + \frac{2}{x^6} - \frac{1}{x^8}$

9 a) 5th term
b) 1st term
c) 5th term

Exercise 2.3

1 ${}_{2n}C_{n-1}$

7 -1 if $n = 1$; 0 if $n > 1$.

9 a) $\binom{n+1}{k} = C_{k-1} + C_k$

b) $\binom{n+2}{k} = C_k + 2C_{k-1} + C_{k-2}$

c) $\binom{n+3}{k} = C_k + 3C_{k-1} + 3C_{k-2} + C_{k-3}$

Exercise 2.4

1 a) $x^2 + y^2 + 4z^2 - 2xy - 4xz + 4yz$

b) $x^3 - 27y^3 + 125z^3 + 15x^2z - 9x^2y + 9y^2x + 45y^2z + 25z^2x - 75z^2y - 90xyz$

c) $16x^4 + 81y^4 + z^4 - 96x^3y + 32x^3z + 8xz^3 - 12yz^3 - 216xy^3 - 108y^3z$
$+ 216x^2y^2 + 24x^2z^2 + 216x^2y^2 + 54y^2z^2$

3 $\left(\frac{-25!}{10!\,11!\,4!}\right) 7^{11} \cdot 16$

Exercise 2.5

1 a) $1 + 5x + 25x^2 \ldots; 5^r$; valid if $|x| < \frac{1}{5}$

b) $1 + 4x + 12x^2 \ldots; 2^r(r+1); |x| < \frac{1}{2}$

c) $1 - 3x + 6x^2 \ldots; \dfrac{(-1)^r(r+1)(r+2)}{2}; |x| < 1$

d) $1 - 8x + 40x^2 \ldots; \dfrac{(r+3)(r+2)(r+1)(-2)^r}{3!}; |x| < \dfrac{1}{2}$

e) $1 + 6x + 6x^2 \ldots; (-1)^r\, 12\, \dfrac{(2r-4)!}{r!\,(r-2)!}$ if $r > 2; |x| < \dfrac{1}{4}$

f) $x - x^3 + x^5 \ldots$; 0 if r is even, $(-1)^s$ if $r = 2s + 1$; $|x| < 1$

3 ${}_{23}C_{14} - 10\,{}_{14}C_5$

7 4

9 a) 3rd and 4th
b) 1st
c) 27th and 28th
d) 1st

11 a) 4

b) 5^6

c) $\sqrt{\frac{2}{3}}$

d) $\frac{1}{3}(179 - 80\sqrt{5})$

13 $-3(1 + x + x^3 + x^4 + x^6 + x^7 + x^9 + x^{10} + x^{12} + x^{13} + x^{15} + x^{16} \ldots)$

$-\frac{5}{4} - \frac{35}{16}x - \frac{125}{64}x^2 + \cdots + (-2 + (-1)^n(3/4)^{n+1} + 1)x^n + \cdots$

Exercise 3.1

1 $(1 + t^2 + t^4 + t^6 + \cdots + t^i)(t + t^3 + t^5 + \cdots + t^j)(1 + t + t^2 + t^3 + \cdots)^{n-2}$

3 ${}_{149}C_{100} - {}_{50}C_1 \cdot {}_{142}C_{93} + {}_{50}C_2 \cdot {}_{135}C_{86} - {}_{50}C_3 \cdot {}_{128}C_{79} + \cdots + {}_{50}C_{14} \cdot {}_{51}C_2$

5 $\binom{14}{10}$

7 $\binom{m+r-1}{m-1}$ if $s + 1 \leqslant r$

$\sum_{j=0}^{k} \binom{m}{j} \binom{m+r-1-j(s+1)}{m-1}$ if $s + 1 \geqslant r$ and k is a positive integer such that $k(s+1) \leqslant r$ but $(k + 1(s+1) > r$.

Exercise 3.2

1 $\sum_{k=0}^{n} {}_nC_k \, k^r \, (-1)^{n-k}$

3 $(1 + t/1! + t^2/2! + t^3/3!)^p \, (1 + t/1! + t^2/2!)^q$

Exercise 3.4

1 $a_n = a_{n-1} + a_{n-2} + a_{n-3} \qquad n > 3;\ a_0 = 1,\ a_1 = 2,\ a_3 = 4$

3 $a\,(n,\ k) = \dfrac{n^2 + 2nk + n + 2k + 2}{2}$

5 $(n^3 + 5n + 6)/6$

Exercise 4.1

5 Let D = deaf mute, B = blind and M = mentally diseased; then $BMD = 25$, $BMD' = 380$, $B'MD = 500$, $BM'D = 82$, $BM'D' = 22{,}980$, $B'M'D = 13{,}585$, $B'MD' = 96{,}478$, $B'M'D' = 28{,}868{,}222$

7 451

Exercise 4.2

1 1485

3 $26!\,(1 - 1 + \frac{1}{2!} - \frac{1}{3!} - \frac{1}{5!} + \cdots + \frac{1}{26!})$

5 dice are different: 1656
dice are indistinguishable: 246

Exercise 4.3

1 4

3 248

5

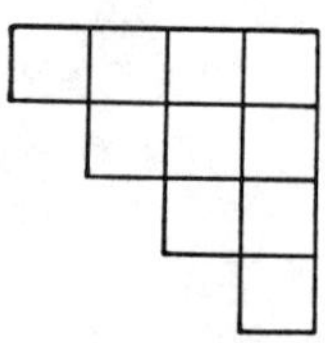

$4! + 3!(10)\,(x - 1) + 2!(25)\,(x - 1)^2 + 15\,(x - 1)^3 + (x - 1)^4$

Exercise 5

1 a) $\frac{2}{5}$

b) $\frac{2}{5}$

3 $\dfrac{{}_{100}C_{15}}{{}_{120}C_{15}}$

5 $\dfrac{2(r-1-k)}{r(r-1)}$

7 Multiply through by n/m. The resulting identity can be interpreted as the probability of getting a black ball for the first time on draw i ($i = 1, 2, \ldots, m-n$) when sampling with replacement from an urn containing n black and $m-n$ white is 1.

9
$$\sum_{j=1}^{n} \frac{\binom{m}{j}\binom{n}{j}}{\binom{m+n}{2j}} - \sum_{i=1}^{n-1}\sum_{j=i+1}^{n} \frac{\binom{m}{i}\binom{n}{i}}{\binom{m+n}{2i}} \frac{\binom{m-i}{j-i}\binom{n-i}{j-i}}{\binom{m+n-2i}{2j-2i}}$$
$$+\sum_{i=1}^{n-2}\sum_{j=i+1}^{n-1}\sum_{k=j+1}^{n} \frac{\binom{m}{i}\binom{n}{i}}{\binom{m+n}{2i}} \frac{\binom{m-i}{j-i}\binom{n-i}{j-i}}{\binom{m+n-2i}{2j-2i}} \frac{\binom{m-j}{k-j}\binom{n-j}{k-j}}{\binom{m+n-2j}{2k-2j}} + \cdots$$
$$+ (-1)^{n-1} \frac{2^n m!\, n!}{(m+n)!}$$

11 Probability that some child breaks 3 or more toys is 53/128.
Probability that the youngest child breaks 3 or more toys is 53/512.

13 a) $\dfrac{{}_{13}C_x \cdot {}_{39}C_{13-x} \cdot {}_{13-x}C_y \cdot {}_{26+x}C_{13-y} \cdot {}_{13-x-y}C_z \cdot {}_{13+x+y}C_{13-z}}{{}_{52}C_{13} \cdot {}_{39}C_{13} \cdot {}_{26}C_{13}} \equiv P(x, y, z, w)$

b) $4P(4, 4, 4, 1)$

c) $24P(5, 4, 3, 1)$

15 $2^{2n-1} \dfrac{(n-1)!\, n!}{(2n)!}$

17 a) $\dfrac{{}_wC_n \cdot {}_bC_{k-n}}{{}_{b+w}C_k}$

b) $\binom{w+b}{k}$

19 a) If $r < n$, prob. is 0;
If $r \geqslant n$, let $A_i \equiv$ car i is empty. Use 21 to find

$$P_0 = 1 - (-1)^n n^{-r} \left[\binom{n}{1} 1^r - \binom{n}{2} 2^r + \binom{n}{3} 3^r - \cdots + (-1)^n \binom{n}{n-1} (n-1)^r\right]$$

$$= (-1)^{n+1} n^{-r} \left[\binom{n}{1} 1^r - \binom{n}{2} 2^r + \cdots + (-1)^n \binom{n}{n-1} (n-1)^r \right.$$

$$\left. + (-1)^{n+1} \binom{n}{n} n^r\right]$$

b) $P_{[n-k]} = n^{-r} \binom{n}{k} \sum_{j=1}^{k} j^r \binom{k}{j} (-1)^{j+k}$

c) $n!\, S(r, n)$

Alternatively, use theorem 3.3.5

Exercise 6.1

1 a) $S = \{a, b, c\}$

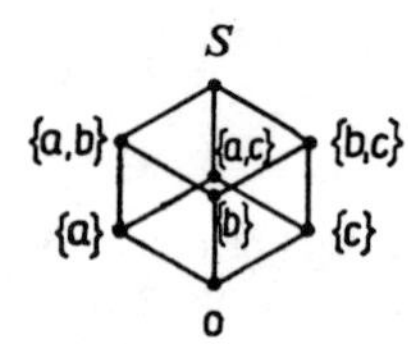

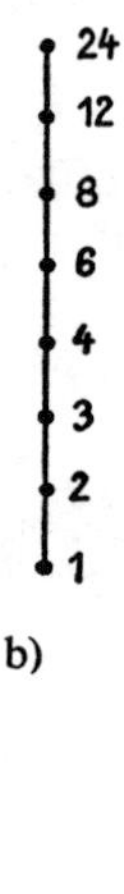

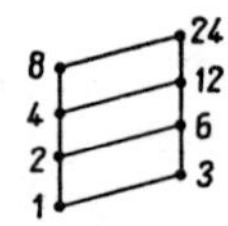

b) c)

Exercise 6.2

1 a) $\mu(m, m) = 1$

$\mu(m, m+1) = -1$

$\mu(m, n) = 0$ if $n \neq m, m+1$

b)

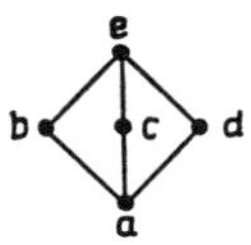

$\mu(x, x) = 1 \qquad x = a, b, c, d, e$

$\mu(a, b) = \mu(a, c) = \mu(a, d) = -1$

$\mu(a, e) = 2$

$\mu(b, e) = \mu(c, e) = \mu(d, e) = -1$

$\mu(x, y) = 0$ otherwise

c)

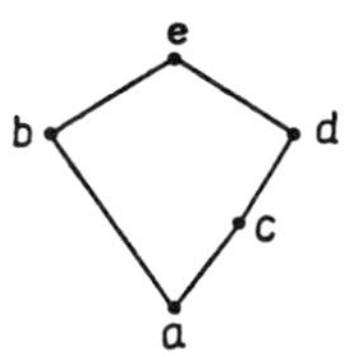

$\mu(x, x) = 1$

$\mu(a, c) = \mu(a, b) = \mu(c, d)$

$= \mu(d, e) = -1$

$\mu(a, e) = 1$

$\mu(x, y) = 0$ otherwise

Exercise 6.3

1 1, 2, 3, 4, 6

3 $\begin{pmatrix} 1 & 2 & 3 & 4 & 5 & 6 \\ 6 & 1 & 2 & 3 & 4 & 5 \end{pmatrix}$, $\begin{pmatrix} 1 & 2 & 3 & 4 & 5 & 6 & 7 \\ 1 & 4 & 2 & 3 & 5 & 7 & 6 \end{pmatrix}$, $\begin{pmatrix} 1 & 2 & 3 & 4 & 5 & 6 & 7 & 8 \\ 2 & 1 & 6 & 3 & 7 & 8 & 4 & 5 \end{pmatrix}$

5 $\{1\}, \{1, A\}, \{1, B\}$ $\{1, AB\}, \{1, A, B, AB\}$

Exercise 6.4

1 Let e = identity, g_1, g_2, g_3, g_4 be typical permutations described in b), c), d) and e) of example 5.2. Burnside's lemma yields $\frac{1}{24}[I(e) + 3I(g_1) + 6I(g_3) + 8I(g_4)] = 2$.

Exercise 6.5

1 (1 2) (3 5) (4 6); (1 2 7 3 6) (4) (5 8); (1 5) (2 4) (3); (1 2 7 8 5 4 6 9 3)

Exercise 6.6

1 a) Let D be the set of 3 chairs and R the set of 3 kinds of paint. Let r, w and b be the weights assigned to the red, white and blue paint respectively. Then the inventory of the set of all functions from D to R is $(r + w + b)^3 = r^3 + w^3 + b^3 + 3r^2b + 3rb^2 + 3w^2b$

$+ 3wb^2 + 3r^2w + 3rw^2 + 6rwb$. This lists all possible ways. To find the total number; set $r = w = b = 1$.

b) Let $D = \{a, b, c, d, e, f, g\}$ be the set of people where a, b belong to one family and c, d to the other. Let $R = \{c_1, c_2, c_3\}$ be the set of 3 cities and let c_1, c_2, c_3 have weights α, β, γ respectively. Then the different ways the people can plan their trips is

$$(\alpha^2 + \beta^2 + \gamma^2)(\alpha + \beta + \gamma)^3.$$

(See lemma 3.) Set $\alpha = \beta = \gamma = 1$.

9 $\frac{1}{2}\left[\frac{1}{n}\sum_{d|n} \phi(d)\, k^{n/d} + k^{q+1}\right]$ for $n = 2q + 1$

$\frac{1}{2}\left[\frac{1}{n}\sum_{d|n} \phi(d)\, k^{n/d} + k^{q}\,\frac{(k + 1)}{2}\right]$ for $n = 2q$

Index